TRANSPORTATION ISSUES, POLICIES AND R&D

HUMAN FACTORS IN AUTOMATED DRIVING SCENARIOS

ASSESSMENT OF RESEARCH, TECHNOLOGY, AND CONCEPTS

TRANSPORTATION ISSUES, POLICIES AND R&D

Additional books in this series can be found on Nova's website
under the Series tab.

Additional e-books in this series can be found on Nova's website
under the e-book tab.

TRANSPORTATION ISSUES, POLICIES AND R&D

HUMAN FACTORS IN AUTOMATED DRIVING SCENARIOS

ASSESSMENT OF RESEARCH, TECHNOLOGY, AND CONCEPTS

JODIE REEVES
EDITOR

nova publishers
New York

For permission to use material from this book please contact us:
Telephone 631-231-7269; Fax 631-231-8175
Web Site: http://www.novapublishers.com

Library of Congress Cataloging-in-Publication Data

ISBN: 978-1-63463-063-4

Published by Nova Science Publishers, Inc. † New York

CONTENTS

PREFACE

This book assesses the research, the technology and the concepts behind automated driving.

Chapter 1 - Although the concept of a fully automated driving system as envisioned under the Automated Highway System program (Congress, 1994) has yet to be realized, technological advancements over the past decade have led to the emergence of advanced driver assistance systems and features such as Adaptive Cruise Control (ACC), collision warning, automatic braking, and lane-keeping assist systems. To date, deployed systems and features have largely been designed to support safe operations rather than to relieve the driver of direct vehicle control. Advancements in driver assistance systems (e.g., ACC and lane-keeping assist) may provide some of the early building blocks for future automated driving systems that assume either partial or full authority from the driver.

Although automated systems offer the promise of increased safety and reduced human error, substantive human factors challenges need to be addressed before these forms of automated systems become a practical reality. These challenges include the potential for negative adaptations occurring through misunderstanding of, misuse of, or overreliance on the system, or changes in attention and distraction from the driving task. Another concern is how an automated system will impact drivers' information-processing capabilities and level of workload, including their willingness to engage in non-driving-related secondary tasks. Automation may also impact a driver's situational awareness – including the ability to perceive critical factors in the environment or to detect system state changes (system failures) – as the driver's role shifts from active vehicle control to passive monitoring of the automated system and environment, and path planning down the road.

Automated vehicles can potentially improve highway safety by supporting or supplementing the driver in different situations. Automation can assist the driver by providing vehicle control during normal driving, or by providing emergency responses in safety-critical situations. The main objective of this research effort is to perform an initial human factors assessment of driver performance and behavior under Level 2 (Combined Function Automation) and Level 3 (Limited Self-Driving Automation) automated driving, with the goal of supporting future research efforts as well as early government policy decisions. As part of this effort, this document summarizing past research, the state of automation technology, and emerging system concepts was generated.

Chapter 2 - The Concepts of Operation document evaluates the functional framework of operations for Level 2 and Level 3 automated vehicle systems. This is done by defining the varying levels of automation, the operator vehicle interactions, and system components; and further, by assessing the automation relevant parameters from a scenario-based analysis stand point. Specific to the "Human Factors Evaluation of Level 2 and Level 3 Automated Driving Concepts" research effort, scenarios and literature are used to identify the range of near- to mid-term production-intent systems such that follow-on research topics with highest impact potential can be identified through commonalities in operational concepts.

In: Human Factors in Automated Driving Scenarios
Editor: Jodie Reeves
ISBN: 978-1-63463-063-4

Chapter 1

HUMAN FACTORS EVALUATION OF LEVEL 2 AND LEVEL 3 AUTOMATED DRIVING CONCEPTS: PAST RESEARCH, STATE OF AUTOMATION TECHNOLOGY, AND EMERGING SYSTEM CONCEPTS*

Tammy E. Trimble, Richard Bishop, Justin F. Morgan and Myra Blanco

ABSTRACT

Within the context of automation Levels 2 and 3, this report documents the proceedings from a literature review of key human factors studies that was performed related to automated vehicle operations. This document expands and updates the results from a prior literature review that was performed for the US DOT. Content within this document reflects the latest research and OEM activity as of June 2013. Studies both directly addressing automated driving, and those relevant to automated driving concepts have been included. Additionally, documents beyond the academic literature, such as articles, summaries, and presentations from original equipment manufacturers and suppliers, have been researched. Information from both United States and international projects and researchers is included. This document also identifies automated-driving relevant databases in support of future research efforts.

SECTION 1. BACKGROUND ON AUTOMATED DRIVING CONCEPTS

Project Background and Purpose

The examination of automation and to what extent various functions should be allocated to either the human or the machine has been a topic of discussion for decades in the field of

* This is an edited, reformatted and augmented version of a National Highway Traffic Safety Administration sponsored document, DOT HS 812 043, issued July 2014.

human factors (Fitts, 1951). Automation can allow for the human to shift from a role of primary responsibility for planning, executing, and monitoring, to one of supervisory control (Sheridan, 1970). In the case of automated driving, depending upon the vehicle's level of automation, automation can allow for some degree of vehicle control to be shifted from the driver to the vehicle. This shift in control represents the potential for enormous safety benefits.

However, this concept is not novel. Automated driving has been discussed in both the scientific and popular literature for decades. General Motors Corporation's (GM's) Futurama exhibit at the 1939 New York World's Fair presented a system for automotive guidance using electrical conductors embedded within the road (O'Toole, 2009, p. 189). However, the driverless car depicted in the World's Fair exhibit has not come to fruition. Instead, a variety of different automation technologies, at different levels of operation and automation, have become commonplace in the vehicle. In fact, the near future will likely include more advanced automation to assist and supplement the driver.

Although the concept of a fully automated driving system as envisioned under the Automated Highway System program (Congress, 1994) has yet to be realized, technological advancements over the past decade have led to the emergence of advanced driver assistance systems and features such as Adaptive Cruise Control (ACC), collision warning, automatic braking, and lane-keeping assist systems. To date, deployed systems and features have largely been designed to support safe operations rather than to relieve the driver of direct vehicle control. Features such as ACC, for example, enable drivers to relinquish partial control over the vehicle in order to increase convenience and safety through the automated management of longitudinal vehicle spacing (i.e., the distance and headway time between a preceding vehicle and the following vehicle is maintained.) Crash avoidance technologies capable of automatically assuming limited control functions under defined situations (i.e., collision-imminent braking, CIB) are also beginning to emerge into the marketplace, as are lane-keeping assist systems that provide drivers with steering torque to help them maintain their position within a lane (i.e., lateral vehicle control). Advancements in driver assistance systems (e.g., ACC and lane-keeping assist) may provide some of the early building blocks for future automated driving systems that assume either partial or full authority from the driver.

Although automated systems offer the promise of increased safety and reduced human error, substantive human factors challenges need to be addressed before these forms of automated systems become a practical reality. These challenges include the potential for negative adaptations occurring through misunderstanding of, misuse of, or overreliance on the system, or changes in attention and distraction from the driving task. Another concern is how an automated system will impact drivers' information-processing capabilities and level of workload, including their willingness to engage in non-driving-related secondary tasks. Automation may also impact a driver's situational awareness – including the ability to perceive critical factors in the environment or to detect system state changes (system failures) – as the driver's role shifts from active vehicle control to passive monitoring of the automated system and environment, and path planning down the road.

Automated vehicles can potentially improve highway safety by supporting or supplementing the driver in different situations. Automation can assist the driver by providing vehicle control during normal driving, or by providing emergency responses in safety-critical situations. The main objective of this research effort is to perform an initial human factors assessment of driver performance and behavior under Level 2 (L2) and Level 3 (L3) automated driving, with the goal of supporting future research efforts as well as early government policy decisions. As part of this effort, this document summarizing past research, the state of automation technology, and emerging system concepts was generated.

Levels of Automation

The National Highway Traffic Safety Administration (NHTSA) has defined five levels of vehicle automation. These definitions are based on the level of driver versus vehicle control. In this taxonomy, as the level of automation increases, the role of the driver shifts from primary control to that of supervisory control. This taxonomy is used to guide all research under the current study, which is focused on L2 and L3 vehicle automation.

Level 0 (L0, No-Automation)

The driver is in complete and sole control of the primary vehicle controls (brake, steering, throttle, and motive power) at all times, and is solely responsible for monitoring the roadway and for safe operation of all vehicle controls. Vehicles that have certain driver support/convenience systems but do not have control authority over steering, braking, or throttle would still be considered "L0" vehicles. Examples include systems that provide only warnings (e.g., forward collision warning (FCW), lane departure warning (LDWS), blind spot monitoring) as well as systems providing automated secondary controls such as wipers, headlights, turn signals, hazard lights, etc. Although a vehicle with vehicle-to-vehicle (V2V) warning technology alone would be at this level, that technology could significantly augment, and could be necessary to fully implement, many of the technologies described below, and is capable of providing warnings in several scenarios where sensors and cameras cannot (e.g., vehicles approaching each other at intersections).

Level 1 (L1, Function-Specific Automation)

Automation at this level involves one or more specific control functions; if multiple functions are automated, they operate independently from each other. The driver has overall control, and is solely responsible for safe operation, but can choose to cede limited authority over a primary control (as in adaptive cruise control), the vehicle can automatically assume limited authority over a primary control (as in electronic stability control), or the automated system can provide added control to aid the driver in certain normal driving or crash-imminent situations (e.g., dynamic brake support in emergencies). The vehicle may have multiple capabilities combining individual driver support and crash avoidance technologies, but does not replace driver vigilance and does not assume driving responsibility from the driver. The vehicle's automated system may assist or augment the driver in operating one of the primary controls – either steering or braking/throttle controls (but not both). As a result, there is no combination of vehicle control systems working in unison that enables the driver to be disengaged from physically operating the vehicle by having his or her hands off the steering wheel AND feet off the pedals at the same time. Examples of function-specific automation systems include: cruise control, automatic braking, and lane keeping.

Level 2 (L2, Combined Function Automation)

This level involves automation of at least two primary control functions designed to work in unison to relieve the driver of control of those functions. Vehicles at this level of automation can utilize shared authority when the driver cedes active primary control in certain limited driving situations. The driver is still responsible for monitoring the roadway and safe operation and is expected to be available for control at all times and on short notice. The system can relinquish control with no advance warning and the driver must be ready to control the vehicle safely. An example of combined functions enabling an L2 system is adaptive cruise control in combination with lane centering. The major distinction between L1 and L2 is that, at L2 in the specific operating conditions for which the system is designed, an automated operating mode is enabled such that the driver is disengaged from physically operating the vehicle by having his or her hands off the steering wheel AND foot off pedal at the same time.

Level 3 (L3, Limited Self-Driving Automation)

Vehicles at this level of automation enable the driver to cede full control of all safety-critical functions under certain traffic or environmental conditions and in those conditions to rely heavily on the vehicle to monitor for changes in those conditions requiring transition back to driver control. The driver is expected to be available for occasional control, but with sufficiently comfortable transition time. The vehicle is designed to ensure safe operation during the automated driving mode. An example would be an automated or self-driving car that can determine when the system is no longer able to support automation, such as from an oncoming construction area, and then signals to the driver to reengage in the driving task, providing the driver with an appropriate amount of transition time to safely regain manual control. The major distinction between L2 and L3 is that at L3, the vehicle is designed so that the driver is not expected to constantly monitor the roadway while driving.

Level 4 (L4, Full Self-Driving Automation)

The vehicle is designed to perform all safety-critical driving functions and monitor roadway conditions for an entire trip. Such a design anticipates that the driver[1] will provide destination or navigation input, but is not expected to be available for control at any time during the trip. This includes both occupied and unoccupied vehicles. By design, safe operation rests solely on the automated vehicle system.

Project Research Questions

The Human Factors Evaluation of Level 2 and Level 3 Automated Driving Concepts project as a whole is centered on six key research questions. The goal is to be able to address each of them at the end of the cumulative research effort, using sound empirical research findings. The research questions are:

1. Can drivers safely interact with and operate vehicles that offer L2 and L3 automation systems; e.g., what is the driver performance profile over length of time in continuous or sustained automation?

2. What are the system performance risks from driver involvement with, and interruption from, secondary tasks (such as portable electronic device use) that could arise when operating L2 or L3 automated vehicle systems?
3. What are the most effective hand-off strategies between the system and the driver, including response to faults/failures?
4. How do drivers engage, disengage, and reengage with the driving task in response to the various states of L2 and L3 automation?
5. How do drivers perform under various operational concepts within L2 and L3 automation, such as systems intended for everyday driving on open roadways in mixed traffic or systems intended for dedicated roadway-vehicle applications (e.g., automated lanes, remote highways)?
6. What are the most effective human-machine interface concepts, guided by human factors best practices, which optimize the safe operation of L2 and L3 systems?

This specific task maps the spectrum of relevant automated vehicle operations and performs a literature review as to key human factors studies. The literature captured within this task includes those which addressed automated driving directly, as well as studies not directly focused on automation but which deal with tasks and scenarios relevant to automated driving that are instructive. Additionally, documents beyond the academic literature have been sought, such as articles, summaries, and presentations from original equipment manufacturers (OEMs) and suppliers, which shed additional light on their roadmap for automation and philosophy for driver role. This effort has captured both American and international projects and research. However, it should be noted that the literature review undertaken in this project is meant to expand upon earlier U.S. Department of Transportation (USDOT) efforts (e.g., Shladover, 2012a, 2012b) and is not meant to be a comprehensive summary of all automation research and efforts. Content within this document reflects the latest research and OEM activity as of June 2013. Within this scope, this document is organized in the following manner:

1. Section 1 presents the project background and purpose, including the NHTSA automated vehicle taxonomy and the project's research questions.
2. Section 2 presents an overview of international programs which have addressed automated driving.
3. Section 3 presents a review of prior human factors studies of vehicle automation, including recent technologies that support automated driving.
4. Section 4 presents lessons learned in the human factors of automation from other domains, such as aviation, rail, and process control.
5. Section 5 presents automation-relevant driving databases that can provide support for future NHTSA and USDOT automation research.
6. Section 6 presents manufacturer approaches to vehicle automation, including past work, current projects, and the role of the driver within an automated vehicle.
7. Section 7 presents a summary of the document, as well as a timeline of vehicle automation.

SECTION 2. INTERNATIONAL PROGRAMS ADDRESSING VEHICLE AUTOMATION

Government-sponsored research programs play a major role in stimulating research and development and policy initiatives in vehicle automation. Key programs and findings related to the current effort are reviewed here. A summary table highlighting key human-factors-related findings is included. More in- depth discussions about many of these projects can be found in the recently released literature reviews of international activity in cooperative vehicle-highway automation systems (Shladover, 2012a; Shladover, 2012b).

Europe

European iMobility Forum Working Group on Vehicle Automation

The iMobility forum is a stakeholder forum supported by the European Commission (EC) and facilitated by ERTICO (an EC Intelligent Transportation Systems, or ITS, stakeholder group). Its purpose is to provide a stakeholder consensus with regard to policy and roadmaps to steer research, development, and deployment (iCar Support, 2013). The forum's vision is:

> Safe, smart and clean mobility with zero accidents, zero delays, no negative impact on the environment and connected and informed citizens, where products and services are affordable and seamless, privacy is respected and security is provided.

For the time period 2011–2020, the iMobility Forum estimates the following potential contributions from ITS (iCarSupport, 2013, Objectives section):

- 30% reduction in the number of fatalities across Europe;
- 30% reduction in the number of seriously injured persons across Europe;
- 15% reduction of road-traffic-related congestion;
- 20% improvement in energy-efficiency; and
- 50% increase in availability of real-time traffic and travel information.

Issues of key interest to the forum include the technical, financial, organizational, and legal framework, including issues related to standardization, certification, liability, privacy, security, driver-vehicle interface (DVI, also referred to as human-machine interface or HMI) for cooperative systems and assisted/partially automated driving. The forum consists of several working groups covering a wide range of topics, including Implementation Road Map, International Cooperation, Research and Innovation, Information and Communication Technologies (ICT) for Clean and Efficient Mobility, Digital Maps, Business Models, Legal Issues, Vulnerable Road Users, and Automation.

At the recent iMobility Forum Automation Working Group meeting, Oonk and Svennson (2013) presented the following as the group's key interest areas:

- Perception (vehicles and road operators): reliable object recognition and tracking, situational awareness, state estimation and prediction, accurate road representation, detection of free space, classification of objects, plug-and-play concepts
- Traffic and transport management: open in-vehicle platform for I2V (infrastructure-to-vehicle) communication and functions; arbitration (negotiation between driver, on-board automation and traffic management center); distributed traffic management and self-organizing concepts (lane assignment, smart ramp metering); determine and advise on the applicable level of automation; supervision of automation by traffic management centers; development of smart logistics corridors with advanced transport management
- Cognition and Human Factors: effects of automated driving over a long period of time; interaction with automation in own vehicle and other road users; mode transitions and mode confusion (when driver is unaware or unsure of current mode) associated with the state of automation; takeover ability and controllability; integration of functions: merging of automated (vehicle-based) sensors with cooperative data acquisition and validation; DVI strategies and concepts

European Commission

Jääskeläinen (2013) stated the EC stance as "Automation is increasingly seen as the only long term option." On an individual level, the highly automated vehicle could provide the driver with assistance in complex driving situations in addition to taking over some driving tasks. On a societal level, automated driving has great potential to significantly improve safety and energy efficiency. Projects conducted on behalf of the EC include Safe Road Trains for the Environment (SARTRE), Highly Automated Vehicles for Intelligent Transport (HAVEit), CityMobil, Highly Automated Driving on Freeways (HAD), REFLECT, and Integrated Human Modelling and Simulation to Support Human Error Risk Analysis of Partially Autonomous Driver Assistance Systems (ISi-PADAS). New projects are expected to start in mid-2013 based on a solicitation seeking proposals for supervised automated driving. Smaller projects are expected to explore business models for the deployment of automation, as well as EC participation in international cooperation with the United States and Japan.

SARTRE

The SARTRE project, funded by the EC, investigated close-headway platooning as a potential early step in automation deployment. The project began in September 2009 and ended in September 2012. As part of the project, researchers demonstrated a road train with three and four vehicles at up to 90 km/h (approximately 56 mi/h) at Hällered proving ground in Sweden. A stated EC budget of €22 M was allocated to co-fund work, with industry developing and demonstrating fault-tolerant and resilient supervised automated driving, and studying the potential use of smart lanes or dedicated lanes for automated vehicles. Specific areas of interest include driver takeover situations, emergency stops which require driving the car to a safe place (thus, rendering the behavior of an automated vehicle predictable for other road users), and socioeconomic, standardization, and legal issues.

Chan (2012) described the SARTRE project, in which Volvo cars and trucks were equipped for close- headway platooning (an L3 system). The project investigated the business model of a human driver being paid to drive the lead vehicle by the occupants in the fully automated following vehicles in the platoon. Prototypes were developed using current production car and truck technologies plus Dedicated Short Range Communications (DSRC) for V2V communications. The cars drove automatically behind the lead vehicle at 85 km/h (approximately 53 mi/h), separated by 5–15 meters. Several hundred kilometers of testing were performed.

Larburu, Sanchez, and Rodriquez (2010) explored drivers' opinions regarding platoon driving through a two-part experiment designed to evaluate intra-platoon gap acceptance and platoon length. This research was conducted as part of the European-Union-funded SARTRE project and used a fixed- base driving simulator. Results indicated that most participants thought that 90 km/h (approximately 56 mi/h) was a comfortable speed for a platoon, that driver information is absolutely necessary, and that an acknowledgement from the driver before starting a maneuver is required during every platoon transition maneuver (e.g., transitions from normal driving to autonomous/automated driving and vice versa; 2010, p. 10). In regard to following distances, in general, people indicated feeling uncomfortable when the intra-platoon gap length was less than 16 m and indicated feeling unsafe under 7 m. Because the recommended gap distance for the driver to feel comfortable is contrary to platoon benefit and safety concepts, the authors concluded that driver training may be necessary to improve trust in the system. When looking at the length of the platoons, approximately 73% of participants indicated that they viewed driving near a platoon of five cars and one leading truck the same as normal driving and did not experience feeling unsafe when driving near the smaller platoon. When platoon size was increased to 15 cars and one leading truck, acceptance was reduced to approximately 55%. With 25 cars and one leading truck platoon size, acceptance reduced to approximately 11%. As such, Larburu et al. concluded that 15 cars and one leading truck should be considered the maximum length for a platoon (2010). However, they note that participants' experience with and trust in platoon systems and backgrounds (i.e., professional drivers), as well as the testing of these concepts in real driving environments, may improve the obtained results.

HAVEit

The EC funded the HAVEit project (project timeline: 2008–2011) to investigate highly automated driving – essentially an L2 system combining ACC and lane keeping. In doing so, the project aimed to develop, validate, and demonstrate important intermediate steps towards automated driving. The project sought to contribute to safety, efficiency, and comfort through three measures (HAVEit, 2012, About HAVEit section, para. 2):

- Design of the task repartition between the driver and co-driving system (termed Advanced Driver Assistance Systems [ADAS] in the HAVEit project) in the joint system;
- Failure-tolerant safe vehicle architecture including advanced redundancy management; and,
- Development and validation of the next generation of ADAS directed towards higher level of automation as compared to the current state of the art.

Within the European HAVEit project (Beutner et al., 2011), several specific demonstrators were developed. A brief overview of the Automated Queue Assistance (AQuA) demonstrator, Joint System Demonstrator (JSD), and Automated Assistance in Roadwork and Congestion (ARC) is provided.

The AQuA demonstrator was developed on a heavy-truck platform and supported a commercial driver in congested traffic by automatically handling the speed and steering control. AQuA aimed to relieve the driver of the monotonous tasks associated with driving a truck in congested traffic situations, i.e., driver underload situations. The level of automated control was continuously adapted across L0 through L2 based on the states of the driver, the vehicle, and the environment. At the highest level of automation, the system autonomously handled steering, acceleration, and braking to keep the vehicle in the correct lateral position in the lane and at a safe distance from the preceding vehicle or at a desired speed. The system operated at speeds between 0 and 130 km/h (approximately 81 mi/h). The developed JSD also provided L0 through L2 automation (Beutner et al., 2011). The vehicle was equipped with warning- and intervention-based active safety, ACC, and a lane-keeping assistance system (LKAS). Steering wheel buttons allowed drivers to choose a specific automation level. Detection of hands on the steering wheel was implemented. The ARC demonstrator was also developed as part of the HAVEit project (Strauss et al., 2010).

The main focus of the ARC demonstrator was driving through a work zone in the highly automated mode (L2). Driving through a work zone could be an overload situation at speed, or an underload situation if congested. The system addressed the possibility that lane lines may not be visible or accurate; therefore, additional objects such as adjacent-lane trucks, beacons, and guide walls were used for guidance. A control algorithm called virtual wall steered the vehicle back into the lane when it got too close to a conventional or unconventional lane border; otherwise, an LKAS provided lane centering and ACC provided longitudinal control. The ARC operated at speeds between 0 and 80 km/h (approximately 50 mi/h).

Through the HAVEit program, a situation-adaptive, optimized task repartition process was defined, taking into account action/status of both driver and co-system to determine the most suitable driving command (Hoeger et al., 2011). The co-system consists of perception, maneuver planning, trajectory computation, and command generation. A key program outcome was the development of a driver engagement process that involved a progressive step-by-step approach to transferring the driving task from the automated co-system to the driver. Within this process, drivers are brought into the loop in advance of critical situations and are provided with a level of automation and assistance determined to be appropriate in supporting the driver during critical situations. The appropriate level of automation and support is determined through driver monitoring/state assessment (via a system that makes a calculation of both driver drowsiness and driver attention levels). When either the driver or the co- system was unable to handle the situation, the automation level was to be changed (e.g., a transition back towards higher driver responsibility or a transition to higher responsibility for the co-system).

As part of the project, a driver state monitor (termed a Driver State Assessment Module, or DSA; Hoeger et al., 2011) was developed to identify drivers' need for automation and to make decisions when the level of automation should be increased or decreased. The DSA used both driver physiological measures and performance measures to derive a model of driver behavior which can be used for detecting driver drowsiness and driver distraction. The authors noted that the available parameters for online driver drowsiness and driver distraction detection are dependent on the current automation level, as some driver performance measures (such as steering variability) will not be available during automated driving (see Hoeger et al., Table 5, p. 107). Techniques deemed as being independent from the current level of automation include reaction time to specific events, indirect measures referring to additional in-vehicle activities (e.g., driver's use of onboard systems), and direct driver monitoring (i.e., the observation of the driver's eye closure and head or gaze direction via camera). Because reaction times require a triggering event, this measure should be used as a continuous measure of the driver's performance abilities if the event is introduced by a secondary task. Noted solutions include the incorporation of a display that explicitly shows the activated level of automation (e.g., the HAVEit automation scale and the color changes in the display) or a hands-on check for the transition to a lower level of automation or a hands-on warning for levels semi- automated and driver assisted. The authors checked different interaction schemes and transition variants. Assessment of user understanding of engage/disengage transitions showed that they adapted well and built accurate mental models. Overall, the HAVEit team concluded that a suitable joint system driver/co-system was successfully developed (Hoeger et al., 2011).

Flemisch, Kaussner, Petermann, Schieben, and Schömig (2010) also reported on HAVEit findings, specifically those related to optimum task distribution. Flemisch et al. provided an overview of three experiments in simulators and one using a test vehicle that focused on the transitions between different modes of assistance and automation, and also on attention-monitoring aspects. The studies evaluated drivers' acceptance of a drowsiness monitor and reactions to an attention monitor, drivers' reactions to two different designs for automation mode transitions, and drivers' evaluations of four prototypes of highly automated driving that included transition schemes similar to those under consideration for HAVEit program vehicles. The results of these studies indicated that the HAVEit concepts for highly automated (L2) driving matched drivers' expectations and received high driver acceptance ratings. These findings suggest that driver monitoring, at least as part of an automated vehicle system, is not likely to meet with driver resistance.

CityMobil

CityMobil was an EC-sponsored research, development, and demonstration project active from 2006 through 2011. This project addressed the integration of automated transport systems in an urban environment (van Dijke & van Schijndel, 2012). A summary of the identified human factors aspects in automated and semi-automated transport systems were identified in project deliverable number 3.2.1 (CityMobil, 2008). Key human factors issues identified within the report related to acceptance and comfort, situational awareness, loss of skill, behavioral adaptation and risk compensation, workload, level of automation and normal transitions, responses to system failures, usability, and guidelines. Additionally, four future scenarios were identified along with associated human factors issues. These four scenarios were assisted vehicles in a town center, urban roads with dual-mode vehicles on an equipped lane, inner city center with advanced fully automated cars, and shared traffic space with automated buses and dual-mode vehicles.

Research and development activities were conducted to identify, address, and, where possible, eliminate barriers blocking the implementation of automated transport systems (e.g., partially automated buses). The demonstration portion of the project included three large-scale implementations of automated transport systems in urban areas and a number of smaller demonstrations. The demonstration activities included (van Dijke & van Schijndel, 2012):

- Heathrow Demonstration: A personal rapid transit (PRT) system featuring small, automatically guided, vehicles for the transport of people called ULTra. The 21 ULTra vehicles carry people from the business car park to a terminal at the airport over a set 3.9-km long route.
- Castellón Demonstration: Partially automated buses were deployed that could be operated both automatically and manually, depending on the road environment.
- Rome Demonstration: A short-distance transport service that uses medium-sized fully automated vehicles to transport people from the parking area to the new Rome exhibition center. Due to nontechnical barriers, the implementation of this demonstration was not completed.
- La Rochelle Demonstration: A temporary demonstration in which advanced fully automated cars (cybercars) following a fixed route provided transport from the harbor ferry to the nearby industrial and living areas. The demonstration ran for two periods of about 3 months.
- Showcases: Demonstrations featuring three advanced fully automated cars (for individual or collective transportation of people or goods) and two advanced city vehicles (i.e., city vehicles which integrated zero or ultra-low pollution mode and driver assistance such as speed adaptation, parking assistance, collision avoidance, and stop-and-go cruise control) were held in five European cities.
- City Studies: City studies were conducted to investigate the advantages of a PRT system (Uppsala, Sweden) and an advanced transport system (Sophia Antipolis, France).

HAD project

The HAD project involved the demonstration of a fully automated vehicle in real traffic on the A9 freeway from Munich to Ingolstadt (Ardelt, Coester, & Kaempchen, 2012). During this drive, the HAD system resulted in safe driving behavior, even during multiple automated lane-change maneuvers. The authors noted that the developed system varied from those presented as part of the U.S. Defense Advanced Research Projects Agency (DARPA) challenges (e.g., Montemerlo et al., 2008; Umson et al., 2008), the Stadtpilot project (e.g., Wille, Saust, & Maurer, 2010a, 2010b) in that sensor locations were restricted, the system was designed for significantly higher velocities (up to 140 km/h, or about 87 mi/h), and lane-change maneuvers did not have to be approved by the driver prior to execution. The test vehicle used for the evaluation was equipped with several sensors (i.e., Differential Global Positioning System (DGPS), radar, camera, laser scanner and ultrasonic), high-precision digital maps, and serial-produced actuators that facilitated electronic control of the steering, brakes, and throttle. The authors found that during the course of the 65 km test drive, the automation displayed an acceptable handling of all occurring traffic situations, including 32 discretionary and mandatory lane changes, without the need for driver approval or intervention.

eLane Study

Toffetti et al. (2009) provided an overview of the human factors issues associated with highly automated vehicles on a dedicated automated vehicle travel lane. These factors include levels of automation, transition of control, loss of driver skill, and responses to automated system errors. In an exploration of these issues, Toffetti et al. conducted a driving simulator experiment to examine the difference between two user interface concepts for a dual-mode vehicle driven both manually and automatically (i.e., L2 automation with automated longitudinal and lateral control). The first interface – labeled the acoustic interface – consisted of visual plus acoustic messages by means of beeps. The second interface – or vocal interface – consisted of visual plus acoustic messages plus vocal messages delivered in English. The two user interfaces were tested when participants had divided attention between the primary driving task and the secondary in-vehicle information systems (IVIS) task. In cases of simulated system failure, the system provided a pre-warning that the system would be deactivated. The warning was followed by a first warning indicating that the system was deactivating and a final warning that an emergency braking maneuver was beginning. Results show that upon the pre-warning step, more drivers with the vocal interface took over control of the car (40% acoustic, 55% vocal); with the acoustic interface, most drivers took control after the first warning (50% acoustic, 36% vocal). No differences existed at the final warning. Further, they found that in cases of system failure, both interfaces resulted in 15% of participants failing to take control of the system in time (i.e., prior to the initiation of an emergency braking maneuver). When compared to the acoustic interface, the vocal interface resulted in faster response times to system failures and greater user preference. The authors concluded that the vocal interface was the recommended interface for the DVI of dual mode vehicles, especially for providing system malfunction warnings. It should be noted that Toffetti and colleagues did not examine other modalities of communication, such as haptic, that may be beneficial for alerting drivers to transition of control events. However, these results do suggest that drivers may prefer and respond better to systems providing vocal/spoken communication. This preference accords with earlier guidelines suggesting (Green, Levison, Paelke, & Serafin, 1994) that non-speech auditory messages such as tones should only be used for alerting, and that voice should be used for more complex messages.

Institute for Transport Studies, Leeds, United Kingdom, Study

Also as part of the CityMobil project, Merat and Jamson (2008) conducted a simulator study to investigate the effects of a highly (L2) automated driving scenario on driver behavior. The researchers compared drivers' responses to critical scenarios during manual and automated driving. Automated driving involved the engagement of lateral and longitudinal controllers, which kept the vehicle in the center of the lane and at a speed of approximately 64 km/h (40 mi/h). Auditory alarms were used to alert drivers of critical situations. When an auditory alarm sounded, drivers were required to regain control of the vehicle. One lateral and three longitudinal critical events were explored:

- A vehicle emerged from a side road and joined the experimental road, driving in front of the lead vehicle;
- An oncoming vehicle turned right to enter a side road, crossing the path of the lead vehicle;

- A set of traffic lights changed to red as they were approached by the lead car; and
- The road was partly blocked by a parked car or a reversing truck.

The findings indicated that drivers' responses to critical events were slower in the automated driving condition as compared to the manual driving condition. Drivers' minimum headway times to the lead car were significantly longer in the manual condition compared to the automated driving condition. Additionally, the results indicated a significant effect of event, with a significantly longer headway on approach to the traffic light event, compared to the other two longitudinal events. Results also indicate that the minimum time to contact with the lead car was significantly different in the manual driving mode, compared to the automated driving mode, with much smaller values seen for the latter condition (1.82 s versus 1.44 s, respectively). Finally, drivers' anticipation of the three longitudinal events was found to be much less in the automated driving condition, with the driver braking just 0.4 s after the lead car braked in the manual condition, compared to 1.9 s after the lead car braked in the automated condition. For the lateral critical event, all but one driver stopped for a gap in the oncoming traffic before overtaking the obstacle. Again, response times were better with manual driving. The average time to contact with the parked car was 4.66 s (SD = 3.55) in the manual driving condition, which reduced to 2.62 s (SD = 0.89) in automated conditions. The authors noted that this difference was statistically significant. They theorized that drivers may have reduced situational awareness during automated driving and an overreliance on the automated system, and concluded that automated driving systems must be designed in such a manner as to keep drivers engaged and in- the-loop so that they are able to respond in a timely and appropriate manner during critical situations.

ISi-PADAS Project

The EC-funded study ISi-PADAS sought to provide a method to support risk-based design and approval of automated driving that focuses on the elimination and mitigation of driver errors through an integrated Driver-Vehicle-Environment modeling approach. A tool-supported risk-based design methodology was introduced to enable evaluation of hazards associated with human error and/or inadequate driver behavior (Transport Research & Innovation Portal, 2013). This methodology focuses on accidents in longitudinal control. Muhrer, Reinprecht, and Vollrath (2012) researched the effects of a forward collision warning and braking system (termed FCW+, also commonly referred to as collision mitigating braking, or CMB) on driving and gaze behavior and engagement in a secondary task. Using a simulator, the researchers had 30 participants take part in a car-following scenario where participants drove with and without FCW+. Results suggest that the use of FCW+ resulted in significantly fewer accidents and earlier reaction times, and was not found to result in stronger involvement in secondary tasks when the system was engaged. Muhrer et al. conclude that partially automated systems such as the FCW+ are necessary because warnings alone cannot prevent accidents.

United Kingdom

Effects of Automation Safety (EASY) Project

Carsten, Lai, Barnard, Jamson, and Merat (2012) explored whether the level of automation that is provided affected driver attention to the road scene and engagement with secondary tasks. Using a high-fidelity motion-based driving simulator, Carsten and colleagues presented drivers with three levels of automation: manual driving, semi-automated (L1) driving with either longitudinal or lateral control provided, and highly automated driving (L3) with both longitudinal and lateral control provided. During the driving tasks, participants were permitted to pay attention to the roadway and traffic or to engage in entertainment or grooming-related tasks. Results suggested that participants were more likely to engage in non-driving tasks as the levels of automation increased. Additionally, in the semi- automated driving scenario, drivers were more likely to engage in secondary tasks when lateral control was in place versus longitudinal control. The authors noted that additional research is needed to further explain differences between levels of automation and differences between longitudinal and lateral control. The authors also suggested that further research is needed to understand how driver attention and interaction with secondary tasks may change over both the duration of the individual exposure to automated driving and over longer periods of time (i.e., in a longitudinal study).

Foot-LITE

Foot-LITE, a U.K.-led consortium (including MIRA Limited, TRW Limited, Auto-txt Limited, Hampshire County Council, the Institute of Advanced Motorists Limited, Ricardo UK Limited, Transport for London, Zettlex Printed Technologies Limited, Transportation Research Group University of Southampton, Brunel University, Transport Operations Research Group Newcastle University, and HW Communications Ltd), aims to bring information on safety and fuel efficiency together on a single, integrated, adaptive interface, providing driver feedback and advice on aspects of safe and green driving styles. As part of the research within this consortium, Birrell and Young (2011) examined automated vehicle driver interface designs using a driving simulator. Two driver interfaces were examined that provided real-time delivery of targeted information to support safe and efficient driving; the impact of these displays on workload, distraction, and driver performance was examined. The first prototype was based on Ecological Interface Design (EID) principles (Burns & Hajdukiewicz, 2004; see also Vicente, 2002 for a review of empirical work by researchers in EID). EID refers to "an interface that has been designed to reflect the constraints of the work environment in a way that is perceptually available to the people who use it" (Burns & Hajdukiewicz, 2004, p. 2). It draws upon the field of ecological psychology, which advocates that human behavior is constrained by their work environments. The three tenets underlying the EID approach are as follows (Burnes & Hajdukiewicz, 2004, p. 4):

1. People must make decisions that are constrained by their work domain.
2. These work domains can be systematically analyzed to determine these constraints.
3. There are design techniques and visualizations that can show these constraints in a way that reduces the need for mental calculation or memory.

The EID design used within this study integrated complex information into a single direct perception display. Safety and ecological information were grouped together on the display, with all parameters presented simultaneously and updating in real time depending on driver input. The second interface design explored by Birrell and Young (2011) was based on a dashboard-type interface developed according to best practices (i.e., the 2008 European Statement of Principles on HMIs for in-vehicle information and communication systems;

European Commission, 2008) and consisted of warning icons (derived from the International Standards Organization [ISO] 2575, 2004) and textual information, with only one parameter shown to the driver at a time. Two different driving scenarios were developed (i.e., Urban, Extra-Urban) which were based on the New European Drive Cycle against which standard emissions data are tested (these standards are maintained by the United National Economic Commission for Europe). The Urban scenario included a city environment, non- divided roadway, a speed limit of 30 mi/h, and eight traffic-light-controlled intersections both with and without pedestrian crossings. The Extra-Urban scenario varied in that it reflected a more urban environment, a divided roadway, varying speed limits (ranging from 40 mi/h to 70 mi/h), and, while free from stop signs and traffic signals, contained other traffic of varying speeds placed in the nearside lane for the driver to navigate. The researchers developed two versions of each scenario in order to minimize learning effects. Findings indicated that the displays did not increase driver workload or adversely affect driver distraction. Additionally, the presence of real-time information was found to decrease mean driving speed in both simple (i.e., the Urban) and complex (i.e., the Extra-Urban) driving scenarios. The results suggested that the EID prototype had a wider effect on speed and acceleration and appeared to have more benefits in terms of driver mental workload and distraction; however, braking was more widely influenced by the second prototype. The authors concluded that, for future iterations, unified displays, adhering to EID principles should be examined for potential use in different levels of automated vehicles.

Intelligent Speed Adaptation (ISA)-UK Project

Carsten et al. (2008) reported on the Intelligent Speed Adaptation (ISA)-UK project. The ISA system incorporates an in-vehicle digital road map onto which speed limits have been coded in combination with a positioning system which can either be a Global Positioning System (GPS) or a GPS-enhanced (with map matching and dead reckoning). Interventions associated with ISA range from advisory (i.e., the driver is informed of the speed limit and of violations), voluntary (i.e., the system is linked to the vehicle controls but the driver can choose when to have the system enabled), or mandatory (i.e., no override is possible). Over several simulator, field trial, and design demonstrations, researchers investigated changes in drivers' behaviors and attitudes towards the ISA system, voluntary usage patterns by road and driver type, differences in truck drivers' and motorcycle drivers' reactions to ISA, the danger associated with certain maneuvers with non-overridable ISA, and the practical implications associated with transferring ISA technologies to other motor vehicles (e.g., trucks, motorcycles).

Carsten et al. (2008) reported on a 6-month field trial evaluation of ISA driving. Seventy-nine drivers participated in 1-month pre- and post-ISA driving for comparison, and 4 months of driving using a car outfitted with thumb-operated, foot-operated, and finger-operated controls located on the steering wheel, accelerator pedal, and central control cluster, respectively. The ISA provided visual (a status information display centrally located in the instrument panel) and auditory (ISA status message giving feedback on system status and activation) information. To account for different potential user patterns and travel patterns, four successive trials were conducted. The trials were conducted in an urban area with private motorists, an urban area with fleet motorists, a rural area with private motorists, and a rural area with fleet motorists. Researchers gathered data on participants' attitudes, acceptance of ISA, and self-reported behavior at various points during the study. For each participant, researchers also conducted four observed drives along a fixed route. The results indicated that the ISA strongly reduced speeding without changing the speed distribution below the speed limit. Additionally, the ISA was most often overridden on roads with speed limits of approximately 113 km/h (70 mi/h), with young and male drivers overriding the system more often than older or female drivers and by more private motorists than fleet drivers. When researchers looked at participant perceptions, no change was noted in perceived behavioral control (i.e., how much control the individual feels he has over his behavior), possibly because participants felt that they had control over their ability to disengage the system. Additionally, participants reported that after using ISA they were less likely to believe that speeding would get them to their destinations more quickly; however, participants were more likely to report that speeding would make them feel good following their ISA experience. Further, participants' satisfaction generally improved over time, physical demand decreased, time pressure increased, feelings of increased risk occurred with ISA when overtaking or driving in fast-moving traffic or on motorways, and situational awareness and awareness of speed limits increased.

A case study in the ISA-UK project, focusing on a delivery truck, was conducted over 9 weeks with 2 weeks for baseline driving, six weeks for ISA testing, and one week for post-ISA testing (Carsten et al., 2008). One driver traveled 6,787 km over the course of the trial. Results indicated that the ISA effectively diminished speeding behavior and speed variability, which was most prominent in the lower speed zones. However, this participant's responses to subjective surveys indicated dissatisfaction and mistrust in the ISA system.

An ISA-UK motorcycle trial consisting of 33 participants was conducted on MIRA Limited's proving ground (Carsten et al., 2008). Because the proving ground did not appear on the digital map used for the project, the onboard digital map was replaced with virtual beacons to locate changes in the speed limits). The DVI on the motorcycle system warned the rider before any intervention occurred. A light- emitting diode (LED) ISA display screen mounted on the handlebars in front of the rider presented speed information, and additional visual cues were given by a pair of red flashing warning LEDs fitted to the left- and right-hand upper edges of the windshield. The windshield warning flashed when power reduction was about to be initiated by the ISA system. Additionally, the rider was provided an auditory alert through earphones connected to the system. Tactile alerts were presented through a vibration unit located underneath the motorcycle's seat. Trials included laps to collect baseline data, advisory ISAs (which provided speed limit information and warning to driver), assisting ISAs (which functioned the same as the advisory ISA but also reduced throttle output when the speed exceeded preset values), and information system trials (which functioned the same as the advisory system but also provided route-related information). Riders indicated that they believed the systems would increase traffic safety; however, these systems were perceived as potentially increasing riders' irritation, stress, and feelings of being controlled, and would decrease the joy of riding. The assisting ISA was viewed less satisfactorily than the advisory or information ISAs. The study also found that only the assisting ISA had a noticeable effect on speed variability.

The ISA-UK project also included a driving simulator study with 26 participants designed to quantify how the presence of a mandatory (with no opt-out function) or voluntary (with an opt-out function) ISA system might affect drivers' overtaking decisions on rural roads (Carsten et al., 2008). Drivers participated in two trials, one with each system. Drivers were told that 50% of the surrounding vehicles were equipped with ISA. Each trial contained 10 overtaking scenarios, half with and half without ISA.

Mental workload scores indicated that drivers rated their driving performance significantly better under the voluntary system than under the mandatory system, for which they reported more system frustration. Drivers rated the ISA systems higher in terms of usefulness than in satisfaction, but the mandatory system was viewed more useful. The voluntary system seemed to have little influence on driver overtaking behavior while the mandatory system reduced drivers' propensity to overtake and compromised the quality of the overtaking maneuvers that occurred.

The Netherlands

The Netherlands Advanced Public Transport Systems (ADPTS) developed an advanced vehicle combining characteristics of a bus, a tram, and an underground vehicle. Related to this effort, Brookhuis and De Waard (2008) explored the consequences associated with the implementation of the advanced vehicle. Twenty-five professional bus drivers completed experimental drives in a driving simulator fitted as a mockup of this vehicle (including the associated controls) that followed the simulated route. The simulator allowed the bus drivers to switch from fully automated control (i.e., metro-type) to semi-automated control (i.e., tram-type with the longitudinal control operated by the driver and the lateral control operated by the automated controller) to fully manual control (i.e., bus-type). The results of the drives led the authors to conclude that a driving license issued for conventional driving should not be applied to driving automated (or semi- automated) vehicles unconditionally, at least not without preparation for driving in automated/semi-automated conditions. They suggested that simulator and on-the-job training and separate licensing procedures in type-approved simulators should be mandatory and noted that the recommended procedure is similar to that used for the training of airplane pilots.

Viti, Hoogendoorn, Alkim, and Bootsma (2008) conducted a large-scale field operational test (FOT) on the public roads in the Netherlands to assess the impact of two advanced driver assistance systems (ACC and LDWS) and to estimate the effects of these systems in traffic flow performance. By focusing on driver adaptation to the ACC system, it was found that drivers did not consider in the same way distance from the leading vehicle in heavily congested conditions as they did in less congested or free-flow conditions. Drivers in medium-dense traffic conditions were found to prefer stable speeds as opposed to stable distance headways. As a result, the ACC system was often deactivated in dense traffic conditions. The authors suggest that the current ACC systems should be seen exclusively as comfort devices, perceived as safety- enhancing, because during dense traffic conditions drivers would be likely to deactivate the ACC systems to instead rely on their own driving skills. The researchers concluded that an ACC system helps drivers to better anticipate and control their decelerations and accelerations. As a result, speed and headway variations are reduced and shockwaves are better stabilized.

Dijksterhuis, Kroiβ, and De Waard (2010) conducted a fixed-base driving simulator experiment to evaluate the effects of providing driving support on lateral control in permanent and adaptive modes. Drivers were provided with driving support in the form of an icon, which reflected the vehicle's lateral position. The icon was presented as a heads-up display (HUD) and was projected on the windshield. This display was chosen to keep the driver's gaze forward. When the driver support was active, drivers were provided with continuous lane width and lateral position information. The simulation consisted of 40-km of roads winding through mainly rural scenery, divided into four main sections of uninterrupted road. The total distance driven was up to approximately 80 km (approximately 50 mi) with speed set to 80 km/h (approximately 50 mi/h). Researchers explored support types (continuously on, off, adaptive support), road width (narrow, or 2.25 m; wide, or 3.00 m), and oncoming traffic density (low, e.g., 6 cars per min for 6 min; high, 40 cars per min for ca. 1 min). The adaptive mode was created by having the display automatically deactivate when lateral position control performance thresholds were exceeded based on general indications of performance during a preceding short time interval (i.e., 30 s). Researchers based two thresholds on deviations from the center lane (i.e., driving in the near-edge-zone for more than 7.5 s during the past 30 s; driving in the over-edge-zone for 3 s during the past 30 s) and one on swerving behavior (i.e., standard deviation of the lateral position set to 22 cm). Participants completed the route and followed auditory instructions during the task. Between road sections, participants were instructed to pull over for a 2-min break during which they completed the System Acceptances Scale (citing Van Der Laan, Heino, & De Waard, 1997) and the Rating Scale Mental Effort (citing Zijlstra, 1993). The researchers found that although not all drivers used the feedback information, particularly in the continuous mode, their results showed positive effects, particularly for the adaptive feedback application. Participants indicated that the adaptive feedback was appreciated for its warning abilities, not its information-providing properties.

Dijksterhuis et al. (2011) also used a driving simulator to determine changes in mental effort in response to manipulations of steering demand. Driving behavior on four narrow lane widths were compared with the normal lane width (3 m) on a standard Dutch two-lane rural road while confronted with oncoming traffic. Steering demand was increased by decreasing the maneuvering space of the driver through a period of high-density oncoming traffic in each lane width section. The researchers found that steering demand factors influenced mental effort expenditure and that using multiple measures contributed to effort assessment. For one of the sections, speed was controlled by the simulator and set to 80 km/h (approximately 50 mi/h) to prevent potential compensatory reactions. Results indicated that for every increasing level of lateral demand, extra effort was mobilized in service of the steering wheel as indicated by a decrease in the standard deviation of the lane position. Additionally, an increase in oncoming traffic was found to be associated with a lateral displacement of the vehicle to a position to the right of the center lane. Further, subjective respondent ratings were only sensitive to different levels of lane width under conditions of high demand. The authors concluded that these findings could be used as a starting point for the development of a driver-monitoring system that assesses mental effort to trigger driving support and to determine the most appropriate types of support.

Building upon these prior efforts, Dijksterhuis et al. (2012) tested the implementation of an adapted driver support system using a simulator. The researchers developed a HUD to provide continuously updated information related to the vehicle's lane width and lane position. In testing this HUD, steering demand was increased through narrowing lane width and increasing density of oncoming traffic, while speed was fixed at 80 km/h (approximately 50 mi/h) in all experimental conditions to prevent compensatory speed reactions. Researchers compared the effects of three support modes: a no- support mode (where the HUD was turned off), a nonadaptive-support mode (the HUD was continuously activated), and an adaptive-support mode. In that adaptive-support mode, the HUD was triggered when the participant exceeded the lateral control performance thresholds (i.e., driving in the near-edge zones for more than 7.5 s, driving in the over-edge zones for more than 3 s, and when the standard deviation of the lane position varied more than 22 cm [approximately 8.65 in]). Support was deactivated when all trigger variables were below their threshold values. The authors found that when the adaptive support was used by participants, their driving behavior improved as compared to the non-adaptive and no-support modes. Additionally, they found that participants preferred the adaptive support mode mainly as a warning signal and tended to ignore non-adaptive feedback (i.e., the HUD); one third of the participants indicated that they ignored the HUD. Based on their findings, the researchers concluded that the effects of a support initiation can be expected, regardless of the support type that is activated.

France

Koustanaï, Cavallo, Delhomme, and Mas (2012) explored the impact that familiarization with an FCW has on driver behavior. A driving simulator study was conducted with three groups of drivers with varying levels of knowledge regarding FCW: familiar with FCW, unfamiliar with FCW, and no contact with FCW. The driving simulator used in the study was equipped with an FCW that triggered a single visual-plus-tone warning when the distance from the lead vehicle could become too short to avoid a collision (as defined by ISO 15632). The FCW's visual interface was displayed on a screen located above the dashboard in front of the driver and consisted of a horizontal bar that lit up in one of three colors: (a) yellow, slower than 50 km/h (approximately 31 mi/h) when the system was inactive; (b) green, faster than 50 km/h; and (c) red, whenever the lead vehicle's distance went below the warning distance. The system sounded a three-beep tone when the bar changed from green to red; the bar remained red as long as the distance was too short. Participants were familiarized with the system prior to the test session. During the test session they encountered five events which elicited no alarms, a nuisance alarm, or useful alarms. They also completed a distraction task. Drivers' perceptions were gathered using 7-point Likert-type scales to measure perceived difficulties of the driving situation, self- assessments of simulator driving, a system assessment, and the mental workload generated by the FCW and the distracting task. The researchers found that familiarization on a simulator involving situations with absent or useless warnings had a beneficial effect in that it increased drivers' understanding of the system and made it seem more trustworthy. As drivers became more familiar with the FCW, driver-system interactions became more effective as demonstrated through the lack of collisions, longer time headways, and better reaction times in most situations. However, although familiarization increased drivers' trust in the FCW, it did not raise system acceptance. Additionally, the system did not eliminate potentially risky behaviors. The authors concluded that practice on a simulator could help drivers learn to properly use a driver assistance system.

Germany

KONVOI

Lenk, Haberstroh, and Wille (2011) and Zlocki (2012) provided descriptions of the 2005 to 2009 German KONVOI project for L3 platooned trucks. Because the project partners did not expect future construction of dedicated truck lanes, the main purpose of the project was to assess the impacts of automated convoys on surrounding traffic (Lenk et al., 2011).

Zlocki (2012) provided an overview of the KONVOI project. Four test trucks were driven on a public roadway. The first truck was driven manually by a driver supported by ACC and an LDWS, with following vehicles operating under an automated platooning system for longitudinal and lateral control. Inter-vehicle gaps were on the order of 10–15 m (Lenk et al., 2011). The platooning system underwent preliminary testing on closed test tracks and closed motorways. Two- and three-vehicle platoons were tested, with the majority of the testing evaluating four-vehicle platoons. In total, over 3,000 km of testing was conducted. The final tests took place on a motorway in real traffic. During the road test, the platoon was accompanied by a preceding and following vehicle and a motorway police escort. Driving maneuvers included coupling, following, and de-coupling. Researchers found comfortable and safe driving in platoons with no amplifications of minor interference and no significant influence on surrounding traffic. Additional results indicated general acceptance towards platoons; however, additional public education regarding platoons is needed, and platoons need to be clearly marked. In terms of operation demand for drivers, simulator studies found driving distances were lower after 2 h of platooning; platooning had no influence on lateral driving behavior (Zlocki, 2012).

Italy

VisLab Intercontinental Autonomous Challenge

The Artificial Vision and Intelligent Systems Laboratory at the University of Parma conducted the VisLab Intercontinental Autonomous Challenge in 2010, traversing a 13,000-km route from Rome to Shanghai over a wide variety of road types, traffic situations, and weather conditions (Broggi et al., 2012). This experiment was aimed at testing autonomous operations in uncontrolled environments over long periods to determine possibly inconsistent behaviors. The overall approach to automation keyed on low-cost and highly integrated sensors. Computer vision was widely used. Seven cameras were installed (five forward and two backward looking), while four laser scanners with different characteristics were placed around the vehicle. Redundancy was provided in the area in front of the vehicle with four laser scanners and five cameras. A four-layer laser scanner was mounted in the middle of the frontal bumper, to frame an area of about 100 degrees; two other laser scanners were mounted on the front-right and front-left corners to frame the front and sides of the vehicle. The fourth laser scanner was mounted on the roof rack and tilted down to frame the ground. Additionally, GPS, Inertial Measurement Unit, and V2V communications were employed.

Automated driving was done in two modes: (1) A leader vehicle defined the route (coarse definition) and host vehicle provided the precise route via on-board sensing; (2) When the leader was not visible, the position was provided by GPS waypoints and position refinement was done by on-board sensing. Vehicle system functions included leader-follower, stop-and-go, waypoint following, vehicle detection, lane detection, obstacle detection, pedestrian detection, and ditch and berms detection (off-road). The vehicles encountered challenging situations, such as road construction with long stretches of off-road tracks, roads with no asphalt and large holes, and low visibility due to dust (Bertozzi et al., 2011). Due to these conditions, 8,244 km were covered in automated mode at an average speed of 38.4 km/h and a maximum speed of 70.9 km/h. The maximum amount of time spent in automated driving in a single day was about 6.5 hours.

Asia

Japan

Ministry of Land, Infrastructure, Tourism, and Transport (MLIT)

The Japanese Ministry of Land, Infrastructure, Tourism, and Transport (MLIT) began investigations of automated vehicles in the early 1990s, and a demonstration of automated driving on public roads was conducted in 1996 (Sakai, 2013). An MLIT study group defined next steps in automated driving research, with findings released in 2012. MLIT is pursuing the Advanced Safety Vehicle – Phase 5 Program, which is focused on crash avoidance (via sensors and vehicle-to-vehicle/infrastructure, or V2X, technology). Road-vehicle communication and V2V communication will play a role in providing traffic information, detailed road maps, and other data. The ultimate objective is automated driving. This group sees public availability of autonomous vehicles in the early 2020s (Sakai, 2013). As reported in *Advanced Safety Vehicle: Realization of Secure and Safe Traffic Society by Harmonizing Humans and Vehicles* (Japan MLIT, n.d.), the Advanced Safety Vehicle – Phase 5 Program has adopted a design philosophy that drivers play the main role in driving safely. Driver-focused work includes driver monitoring, driver acceptance, smooth transfer of control, and avoiding system designs which induce over-trust and/or overreliance. As part of this effort, the impacts of automated driving are also being evaluated; identified impact areas include institutional, social acceptance, need for road infrastructure, and socioeconomic impacts.

Several projects are currently underway. For example, researchers are focusing on an Emergency Stopping Assistant (ESA) prototype. In the case of an incapacitated driver, a highly automated driving mode will take control and bring the car to a safe stop at the side of the road (MLIT, n.d.). Additionally, a pilot project is underway to counter congestion at “sag sections” of the roadway, i.e. a slight downhill followed by a moderate incline in which vehicles tend to slow down without the driver’s intention. The sag congestion countermeasures center on the use of cooperative ACC (CACC) and speed advisories generated by the central traffic management center broadcast to vehicles in specific sections of the road segment to maintain appropriate inter-vehicle distance (MLIT, n.d.).

As part of the Advanced cruise-assist Highway System (AHS) project in Japan, Hirai et al. (2007) developed a safety countermeasure for use on merge ramps. This countermeasure provided merging assistance through a cooperative vehicle-highway system based on a DSRC roadside wireless unit communicating with onboard ITS equipment (5.8 GHz). The system was designed so that the AHS detects a merging vehicle and prompts a warning to the vehicle on the main roadway via road-to- vehicle communication (i.e., a DSRC roadside wireless unit and wireless communication system). Through warning sounds, audio, and video warning, drivers would be alerted to either a merging vehicle or the existence of a merging section, the latter of which is provided regardless of the presence of a merging vehicle. To allow time for transmission as well as driver decision and reflex, warning information was provided at least 4 s before the tip of the gore point (i.e., the tip of the triangular piece of land found where roads merge or split). Driver receptivity was determined through a simulator trial and test course trial conducted using the National Institute for Land and Infrastructure Management test course. The results indicated that drivers were receptive to the system and that it was appropriate to display the information on the screens of car navigation systems. Additionally, an expressway trial was conducted to assess service performance and effectiveness. The expressway trial involved 35 trial subjects driving past the merging section five times, for a total of 175 events. The results indicated that the service reduced aggressive merging. Further, they found that although drivers were generally receptive to the service, the implementation of the service was accompanied by an increase in drivers' perceptions of situational dangers.

Ministry of Economy, Trade, and Industry (METI)

As reported in Shladover (2012), the Japanese Ministry of Economy, Trade, and Industry (METI) has a long history of involvement with vehicle automation technology. Initial projects were based on automated vehicles which operated independently and with no cooperation from other vehicles or the infrastructure. In 2008, METI began the Development of Energy-saving Intelligent Transportation Systems Technology (Energy ITS) project focusing on automated truck platooning based on V2V cooperation, with a key aim to reduce emissions. The 5-year project is funded at $12 million per year. The work is being conducted by a variety of university researchers and is led by the Japan Automobile Research Institute. As part of the Energy ITS project, researchers explored freight operators' expectations surrounding the implementation of truck platoons in Japan (Tsugawa, 2012). An automated truck platoon was developed that used lane marker detection for lateral control (i.e., passive and active computer vision) and gap measurement for longitudinal control (i.e., radar, laser scanner, inter-vehicle communications). Initially, a platoon of three heavy trucks (25 T) traveling at 80 km/h (approximately 50 mi/h) with 10-m between-vehicle gaps was demonstrated. Later, a platoon of three heavy trucks and one light truck at 80 km/h with 4-m between-vehicle gaps was demonstrated. During the demonstration, platoon engaging, lane changing, and passenger car cut-in maneuvers were performed. Tsugawa noted several technical issues associated with platoons, including the need for a passive safety device when inter-vehicle gaps are small (i.e., under 4-m platooning) and an HMI (located on the dashboard and on the back of the leader) that provides information to drivers in the following trucks and also to drivers around the platoon.

Market research was also conducted. Freight operators were invited on a ride-along demonstration in one of the project's automated platoons (Tsugawa, 2012). They were then asked about their expectations regarding the potential benefits of automated platooning. The majority of freight operators indicated that they expected platooning to result in energy savings (91%), congestion reduction (79%), load reduction (73%), and high company brand image (60%). In regards to safety, 39% indicated that platooning would result in safe driving while 60% indicated that safety benefits were unknown or not expected. When asked about workload reduction, 67% expressed that they did not expect platooning to result in workload reductions, while 15% indicated that they expected workload reductions; 18% were unsure of workload reductions.

South Korea

Research contributing to the advancement of automated vehicles has been sponsored by the Program of the Ministry of Education Science and Technology (MEST) and the Green

Drive Research Program of the Ministry of Knowledge Economy (MKE). Kim and Son explored the in-vehicle driving workload for older drivers through an on-road assessment of advanced driver assistance systems that assist in safe driving, and IVIS, which offers many types of information to drivers (2011). Kim and Son investigated the effect of age-related workload difference through the examination of five driving tasks (manual only, manual primarily, visual only, visual primarily, and visual-manual) and three age groups: younger (20–29 years of age), middle-aged (40–49 years of age), and older (60–69 years of age). They collected data from 40 drivers who drove in real vehicles under actual road conditions. During the city drive, participants were asked to complete five tasks which could be classified as manual only (activating the turn signal); primarily manual (adjusting the radio); visual only (reading the speedometer); primarily visual (changing the radio station); and combined visual and manual (setting the correct temperature in a menu-driven climate control display). After the drive, participants were asked to complete the National Aeronautics and Space Administration (NASA) Task Load Index (NASA-TLX; Hart & Staveland, 1988) questionnaire. Results showed that older drivers needed more time to complete tasks than did younger drivers and tend to exceed distraction guidelines (i.e., the standards for preventing distraction due to the operation of information devices (see Bischoff, 2007; Li et al., 2002) when completing relatively complicated tasks. Results indicated that older drivers spent 13.08 s and 15.60 s to complete primarily visual and combined visual and manual tasks, respectively. They noted that these findings suggest older drivers may be more likely to be involved in traffic accidents when operating these devices while driving. The authors concluded that, when designing information devices for intelligent vehicles, designers need to take these facts into account and to include methods of reducing distraction for older drivers.

SECTION 3. PRIOR HUMAN FACTORS STUDIES OF VEHICLE AUTOMATION CONCEPTS

Levels of Automation and Automation Taxonomies

Several attempts have been made to create taxonomies, both general and application-specific, of automation. In discussing vehicle levels of automation, it may be helpful to examine prior efforts to produce automation taxonomies and how they may be compared and applied to the driving domain. While some discrepancies may exist when applying taxonomies intended for computer control systems, air traffic control support, or similar domains, to the driving task, understanding the commonalities between the efforts can help lend better understanding of the current state, and future challenges, of vehicle automation.

Perhaps the most widely used taxonomy was presented by Sheridan and Verplank (1978; Table 3-1), who defined 10 levels of automation that account for the locus of control (human or automation) and how information is presented to the human. This taxonomy provides a range of automation levels where the human is responsible (generally, below L5), where the automation holds responsibility (generally, above L7), and automation levels where the system is collaborative (generally, between L5 and L7; Spiessl et al., 2011). This taxonomy was later expanded upon by Parasuraman, Sheridan, and Wickens (2000) to include four phases of processing (perception, analysis, decision-making, and execution) for each level.

Table 3-1. Sheridan and Verplank's (1978) Levels of Automation

Level	Description
1	Human does the whole job up to the point of turning it over to the computer to implement.
2	Computer helps by determining the options.
3	Computer helps to determine options and suggests one, which human need not follow.
4	Computer selects action and human may or may not do it.
5	Computer selects action and implements it if human approves.
6	Computer selects action, informs human in plenty of time to stop it.
7	Computer does whole job and necessarily tells human what it did.
8	Computer does whole job and tells human what it did only if human explicitly asks.
9	Computer does whole job and decides what the human should be told.
10	Computer does whole job if it decides it should be done, and if so, tells human, if it decides that the human should be told.

Table 3-2. Riley's (1989) Taxonomy of Automation Levels

Level of Autonomy	**Level of Intelligence: Raw Data**	**Level of Intelligence: Procedural**	**Level of Intelligence: Context Responsive**	**Level of Intelligence: Personalized**	**Level of Intelligence: Inferred Intent Responsive**	**Level of Intelligence: Operator State Responsive**	**Level of Intelligence: Operator Predictive**
None							
Information Fuser							
Simple Aid							
Advisor							
Interactive Advisor							
Servant							
Assistant							
Associate							
Partner							
Supervisor							
Autonomous							

Many parallels may be drawn between the Sheridan et al. (1978) levels of automation and the definitions of vehicle automation levels proposed by NHTSA (see Section 1). Examples of L0 (manual driving) automation can be compared to Sheridan's L3. An L0 automated component, such as FCW, determines an option (in this example, whether a forward crash threat is present) and informs the driver. Continuing this parallel, higher levels of vehicle automation follow higher levels in the Sheridan taxonomy. CIB (an example of a L1 automated component) can be compared to L6 and L7 of Sheridan's taxonomy, as the system is determining the presence of a threat and implementing a response while keeping the driver informed.

Riley (1989; Table 3-2) presented a taxonomy of automation levels as applied to a mixed-initiative human-machine system. The Riley taxonomy presents automation as two crossed factors: levels of intelligence and levels of autonomy. Levels of intelligence increase across seven levels, from raw data (no processing) to operator predictive (with the automation anticipating actions). Levels of autonomy range across 12 levels, from none to autonomous. Riley's taxonomy allows for any system's automation state to be described as a function of these two factors.

The Riley (1989) taxonomy does not provide for automation responsibility until the Assistant and higher levels; the automation component is responsible for communicating information back to the user at these lower levels of automation. At higher levels, greater responsibility is present in the automation; the Partner level allows for the automation to override the operator (and vice-versa). These levels of automation may also be compared to the NHTSA levels of automation. Many in- vehicle information systems and similar displays can be compared to the Information Fuser level. FCW, as a Level 0 component, may be compared to the Simple Aids. CIB may be compared to the Servant level. NHTSA Levels 2 and 3 may likewise compare to the Associate level. L2 automation provides for lateral and longitudinal control with the expectation of driver monitoring, and L3 provides for the same automated control without the expectation of driver monitoring. The Associate level allows for autonomous operation without explicit permissions, but with the understanding that the operator may override the automation at any point.

Table 3-3. Endsley and Kaber's (1999) Levels of Automation

Level	Description
Manual Control	The human performs all tasks including monitoring the state of the system, generating performance options, selecting the option to perform (decision-making), and physically implementing it.
Action Support	The system assists the operator with performance of the selected action, although some human control actions are required.
Batch Processing	Although the human generates and selects the options to be performed, they then are turned over to the system to be carried out automatically.
Shared Control	Both the human and the computer generate possible decision options. The human still retains full control over the selection of which option to implement; however, carrying out the actions is shared between the human and the system.
Decision Support	The computer generates a list of decision options that the human can select from or the operator may generate his or her own options. Once the human has selected an option, it is turned over to the computer to implement.
Blended Decision-Making	The computer generates a list of decision options that it selects from and carries out if the human consents. The human may approve of the computer's selected option or select one from among those generated by the computer or the operator. The computer will then carry out the selected action.
Rigid System	Representative of a system that presents only a limited set of actions to the operator. The operator's role is to select from among this set. He or she may not generate any other options.

Table 3-3. (Continued)

Level	Description
Automated Decision-Making	The system selects the best option to implement and carries out that action, based upon a list of alternatives it generates (augmented by alternatives suggested by the human operator).
Supervisory Control	The system generates options, selects the option to implement, and carries out that action. The human mainly monitors the system and intervenes if necessary.
Full Automation	The system carries out all actions. The human is completely out of the control loop and cannot intervene.

Table 3-4. Gasser and Westhoff's (2012) Description and Categorization of Automated Driving Functions Drawn from the Report of the BASt Expert Group

Level	Task of the Driver According to Automation Level	Exemplary Systems
Fully Automated	The system takes over lateral and longitudinal control completely within the individual specification of the application. • The driver *need not* monitor the system • Before the specified limits of the application are reached, the system requests the driver to take over with sufficient time buffer. • In absence of a takeover, the system will return to the minimal risk condition by itself. • All system limits are detected by the system, the system is capable of returning to the minimum risk condition in all situations.	Motorway-Pilot • Automatic longitudinal and lateral control. On motorways up to an upper speed limit. The driver need not monitor. In case the driver does not react to a takeover request, the system will brake down to a standstill.
Highly Automated	The system takes over lateral and longitudinal control for a certain amount of time in specific situations. • The driver *need not* permanently monitor the system as long as it is active. • If necessary, the driver is requested to take over control by the system with a certain time buffer. • All system limits are detected by the system. The system is not capable of re-establishing the minimal risk condition from every initial state.	Motorway Chauffeur • Automatic longitudinal and lateral control. On motorways up to an upper speed limit. The driver need not permanently monitor. In case of a takeover request, the driver must react within a certain time buffer.
Partially Automated	The system takes over lateral and longitudinal control (for a certain amount of time and/or in specific situation). • The driver must permanently monitor the system. • The driver must at any time be prepared to take over complete control of the vehicle.	Motorway Assistant • Automatic longitudinal and lateral control. On motorways up to an upper speed limit. Permanently monitored by driver.

Level	Task of the Driver According to Automation Level	Exemplary Systems
Assisted	The driver continuously accomplishes either lateral or longitudinal control. The other/remaining task is accomplished by the automating system to a certain level. • The driver must permanently monitor the system. • The driver must at any time be prepared to take over complete control of the vehicle.	Adaptive Cruise Control • Longitudinal control with adaptive distance and speed control. Parking assistance: Lateral control is accomplished by the parking assistance.
Driver Only	The driver continuously (throughout the trip) accomplishes longitudinal (accelerating/braking) and lateral (steering) control.	No (driver assistance) system active that intervenes into longitudinal and lateral control.

Endsley and Kaber (1999; see Endsley's earlier [1987] taxonomy for automation in a decision-support environment) proposed an automation taxonomy for use in real-time control tasks. Among the important aspects describing real-time control tasks that may be related to the driving task are multiple simultaneous tasks (with different relevance to performance and goals) requiring the operator's attention and high-task demands with limited time resources. This taxonomy is presented in Table 3-3.

Here it may be helpful to continue with the examples of FCW and CIB. FCW may be viewed as falling under the Action Support level, as it simply provides the driver with assistance in performing the driving task. CIB, providing for an automated braking response in the absence of driver intervention, may be viewed as a Supervisory Control task. However, the comparison is weakened as the Endsley and Kaber (1999) taxonomy assumes the driver is monitoring the roadway and automation performance.

The Bundesanstalt für Straßenwesen (BASt) leads the German working group on automated vehicles dealing with the Vienna Convention and German laws and regulations that may be a barrier for introducing automated vehicles in Germany. Gasser and Westhoff (2012) provided insight on the BASt efforts to define current and future automation and legal issues. The BASt expert group identified five levels of automation based on the degree of automation: full automation, high automation, partial automation, driver assistance, and driver only. The expert group further clarified these definitions by incorporating the added dimensions of driver tasks according to level, speed range, and utilization time (Table 3-4).

Overall, each of the aforementioned levels of automation taxonomies provides a means to describe the human-automation interaction. Each of the taxonomies provides some features, dimensions, or concepts that are applicable to the examination of vehicle automation. Sheridan and Verplank's (1978) description of the locus of control (varying from the human, to collaborative, to automation), Riley's (1989) description of automation intelligence, and Endsley and Kaber's (1999) concept of decision support are all highly applicable to the driving domain and should be considered when describing vehicle automation. The BASt definitions were designed with vehicle automation in mind and have attempted to account for a number of dimensions specifically associated with the driving task (Gasser & Westhoff, 2012).

Overview Studies

Several researchers have provided comprehensive overviews or reviews of the issues associated with human-automation interactions. For example, Lee and See (2004) provided a comprehensive review of trust in automation which considers trust from a variety of perspectives and considers how automation context and cognition processes affect the appropriateness of trust. Sheridan and Parasuraman (2005) outlined recent research and challenges in the area. Additionally, Sheridan and Nadler (2006) reviewed 37 accidents that occurred in aviation, other vehicles, process control, and other complex systems where human-automation interaction was involved. They noted the implications about causality with respect to design, procedures, management, and training. Nof (2009) edited the *Springer Handbook of Automation*, within which are several chapters relevant to this discussion. For example, Lee and Seppelt (2009) provided an overview of the human factors in automation design, Nakanishi (2009) discussed major initiatives and technologies being developed in the United States and how these developments interact with drivers, and Ollero and Castaño (2009) focused on the control and navigation of automated vehicles and human interactions at different levels. Martens and Jenssen explored behavioral adaptation and acceptance in the *Handbook of Intelligent Vehicles* (2012).

Further, several researchers provide overviews of various human factors issues associated with automation. Stanton and Young (1998) presented several issues relevant to vehicle automation that they noted should be considered in future empirical studies. These issues include locus of control (i.e., the extent to which removal of control from the driver affects vehicle performance), the trust the driver has in the automated system, the situational awareness of the driver as to the operational status of the technological system and the driving context, the mental representation that the driver develops of the automated system, and the mental and physical workload associated with automation, feedback, and driver stress and its implications for automation.

Saffarian, de Winter, and Happee (2012) explored the challenges associated with automated driving systems from a human factors perspective and identified DVI needs for automated vehicles and proposed available solutions. Saffarian et al. (2012) proposed a number of available solutions to address these challenges, including shared control, adaptive automation, use of an information portal, and new training methods. Current definitions of automated driving (see Section 1) include some provisions for accommodating the challenges that Saffarian and colleagues noted. Further, and more focused, research on these topics could be beneficial.

Parasuraman and Manzey (2010), through a review of empirical studies of complacency and bias in human interaction with automated and decision support systems, concluded that complacency and automation bias results from the dynamic interaction of personal, situational, and automation-related characteristics. Parasuraman and Wickens (2008) reviewed empirical studies of human-automation interaction and their implications for automation design. They found that automation applied to information analysis or decision-making functions leads to differential system performance benefits and costs that need to be considered when choosing appropriate levels and stages of automation. Additionally, they found that human user dependence on automated alerts and advisories reflected two components of operator trust – reliance and compliance – which, in turn, are determined by the threshold designers have used to balance automation misses and false alarms. Further, they found that adaptive automation can provide additional benefits in balancing workload and maintaining the user's situational awareness. Parasuraman, Sheridan, and Wickens (2008) also reviewed the empirical evidence associated with situation awareness, mental workload, and trust in automation. Through this review, they concluded that situation awareness, mental workload, and trust are viable constructs that are valuable for understanding and predicting human-machine performance in complex situations.

Additionally, Lee (2008) noted that the addition of new technologies introduces new vulnerabilities (e.g., distraction) into the driving process. New technologies, such as those associated with driver- assistance technologies, may mitigate distractions and improve safety; however, imperfect technologies that automate driving rather than augmenting driver capabilities may be rejected or misused by drivers. Therefore, the greatest safety benefits from current and forthcoming automated driving concepts may come from augmenting the driver's capabilities and performance with automation rather than attempting to replace the driver with an automated system. Cognitive engineering principles, which enhance drivers' self-awareness and awareness of the potential distractions associated with technology, can be used to improve the safety and performance of complex systems (Lee, 2008). Inagaki (2010) also provides a review of the issues associated with the development of ADAS from a joint cognitive systems approach. Specifically, the issues of authority and responsibility and of overtrust and overreliance were discussed.

The following sections will explore in detail the human factors-related studies associated with automated driving and in support of automation driving. Research supporting automated driving includes those human factors studies associated with the development and implementation of IVISs and DVIs (also referred to as HMIs), ACC, FCW, lane maintenance, and LDWS, and connected vehicle initiatives.

Studies of Automated Driving

This section provides an overview of the studies of automated driving. A diverse group of researchers are actively developing automated vehicle capability for all road operations (including unpaved roads). These are briefly summarized here. Key efforts in this area included work conducted by the U.S. Army (Theisen, 2011; Schoenherr, 2009), Israel's Ministry of Defense (Main, 2013), private industry (Rio Tinto, 2012), the Oxford Mobile Robotics Group (2013; Lee, 2013), and the simulator study conducted by Neubauer et al. (2011).

As described in Theisen (2011), the U.S. Army is sponsoring development of the Autonomous Mobility Appliqué System, a program designed to retrofit existing military trucks with a range of systems, from active safety to full automation. The intent is for the vehicles to operate on any road type as well as off-road. BAE Systems (a producer of military heavy trucks) is the vehicle industry participant. Additionally, automated convoy systems are being examined, such as the Convoy Active Safety Technology (CAST) system. CAST is intended as a low-cost automated following system for tactical wheeled vehicles (Schoenherr, 2009). The CAST system maintains an inter-vehicle gap, adjusting vehicle speed and shifting gears as needed. The vehicles are capable of maneuvering around corners and navigating difficult terrain. Obstacle detection and avoidance is implemented. Following vehicles can assume command of the convoy automatically if the lead vehicle becomes inoperable. This system has achieved more than 100 mi of continuous autonomy with an approximate 80-km/h (50- mi/h) capability on paved roads and an approximate 56-km/h (35-mi/h) capability on dirt roads.

Theisen (2011) provided additional CAST findings. During testing, soldiers wore an electroencephalography (EEG) cap to record brain waves related to their reactions to driving autonomously. Theisen noted that there were no reported adverse physiological effects (e.g., motion sickness issues) associated with the system; participants reported significantly less fatigue, increased ease of convoy execution, and increased numbers of threats detected. The CAST system showed 150% improvement in gap distance maintenance and an 85% improvement in panic stopping distance over a non-automated system. Convoys operating using CAST demonstrated successful daylight driving at 85 km/h (approximately 53 mi/h) and successful blackout driving at 70 km/h (approximately 43 mi/h; Theisen, 2011). This reduction in fatigue may have beneficial effects for eventual production L2 and L3 systems. L2 and L3 automated vehicles have the potential to aid in overall driving performance by ensuring vehicle heading while allowing the operator to rest or perform other tasks. While the number of inputs, tasks, and responsibilities for the driver are not entirely in their direct control, L2/L3 automated vehicles can assist through a decrease in total driver workload. This decrease can, in turn, decrease fatigue, add convenience, and improve overall performance. Further research in this topic is warranted.

In similar autonomous vehicle research, Main (2013) described automated vehicles developed by Israel's Ministry of Defense. Since 2008, approximately eight automated cars have been patrolling Israeli borders. The vehicles are produced by the G-NIUS company. They use cameras, radar, and laser technology for situational awareness. The article noted that these vehicles have taken the place of some soldiers on the front lines, possibly preventing them from facing gunfire or confrontations.

Rio Tinto (2012) noted that the Rio Tinto Yandicoogina mine in Western Australia is now operating a fleet of 10 Komatsu driverless haul trucks. Rio Tinto has been testing the Komatsu Autonomous Haulage System since 2008; during trials, the autonomous haulage technology demonstrated benefits in health, safety, and productivity. A company spokesman was quoted as saying Rio Tinto eventually plans to deploy 150 driverless trucks, making them the world's largest owner and operator of these vehicles.

Lee (2013) describes the Oxford University RobotCar project. The base vehicle is a Nissan LEAF electric car. The car is equipped with an off-the-shelf computer and is connected to machine-vision cameras and laser sensors around the car's body. As opposed to GPS, the RobotCar relies upon machine-learning technologies to build and calibrate mathematical models which explain the robot's view of the world in terms of prior experience (training), prior knowledge (aerial images, road plans, semantics), and automatically generated Web queries. The car develops situational awareness made up of both static and dynamic environmental features. If an obstacle is detected, the vehicle comes to a controlled stop and waits until the obstacle has moved out of the way before accelerating and continuing its journey. The HMI is provided via an in-car tablet computer. When the car successfully recognizes the driving environment, it prompts the driver and offers to take over driving. Tapping the brake pedal returns control to the human driver (Lee, 2013). The researchers believe that this approach will enable a low-cost approach to future automated vehicles. Testing is occurring in Oxfordshire at a light industrial site with roads and road markings (Lee, 2013).

Neubauer et al. (2011) used a driving simulator to assess the impact on fatigue, stress, and workload of driver-initiated full vehicle automation (i.e., L3; drivers were not required to keep their hands and feet on the controls). They noted that previous research demonstrated that full automation induced a state of "passive fatigue" associated with the loss of alertness. This state can be seen as similar to a vigilance decrement (Parasuraman & Davies, 1977). Neubauer et al. asked participants to drive in either an automation-optional or non-automation 35-min, monotonous simulated drive using a driving simulator. Participants had the option of initiating automation at their discretion for 5-min blocks throughout the drive. Researchers assessed subjective stress states before and after the drive using the Dundee Stress State Questionnaire (DSSQ; Matthews et al., 2002). During the automation- optional condition, researchers tested whether choosing automation was related to subjective state. Specifically, the automation-optional condition was used to determine whether those drivers in the automation-optional condition showed higher task engagement than those in the non-automation option. Additionally, researchers explored whether the use of automation at a time of the driver's own choosing would act as a rest break and relieve driver fatigue. In the last 5 min of the drive, in both the automation-optional and the manual condition, drivers had to manually respond to an emergency event. The results suggest that optional, driver-controlled automation (i.e., the automation available as part of the automation-optional condition) appeared to pose the same dangers to task engagement and alertness as externally initiated automation, which may leave drivers vulnerable to persisting fatigue (and the associated reduction in situational awareness) when normal vehicle control is restored. As such, Neubauer and colleagues concluded that an automated systems' impact on driver fatigue needs to be evaluated, and solutions which will maintain driver engagement and address the vulnerabilities of fatigue-prone drivers should be sought.

Research Approaches Supporting Automated Driving

While much of the work examining automated driving has either been related to specific development projects (e.g., the HAVEit project, described above) or more exploratory in nature, more extensive work performed has examined what may be considered underlying technologies that could be considered part of an L2 or L3 automated driving system. These efforts include the DVI design and evaluative approaches of IVIS, especially with regard to such systems' effect on driver workload. Further efforts are associated with the development of ACC and FCW systems, lane maintenance and LDWS, and connected vehicle initiatives. Human factors issues related to these enterprises are discussed in the following sections.

IVIS and Associated DVI Issues

This section describes those research activities that occurred in conjunction with the development of IVIS and related DVI systems. Studies within this section explored the impacts of ADAS on driver workload, while others presented various methods for determining workload. Further, several discussions provided insight as to the design of multi-modal DVIs. This section concludes with the discussion of a DVI designed to adjust to a driver's expectations and preferences. Under concepts of shared authority (see the discussion of levels of automation and associated taxonomies, earlier in this section) for automated vehicles, designers of automated systems should consider the tasks assigned to the driver and to the vehicle. Potential exists for a situation where both the human driver and the automated system simultaneously perceive the other to be the ultimate controlling authority. Thus, DVI should be designed with the goal of preventing this mode confusion. More direct research could qualitatively and quantitatively assess this situation as it pertains to L2 and L3 automated vehicles.

Davidse, Hagenzieker, van Wolffelaar, and Brouwer (2009) examined the extent to which driving performance of 10 older (70- to 88-year old) and younger (30- to 50-year old) drivers improved as a result of support by a driver assistance system. The ADAS tested provided drivers with information on right-of-way regulations and one-way streets, warned drivers when they were approaching an intersection at which sight on the crossing street was obstructed by buildings, and indicated if it was safe for drivers to join or cross traffic streams at busy intersections. The system was tested using a driving simulator and evaluated in terms of effects on workload and safety performance. Findings indicated that the messages that informed drivers of right-of-way regulation, obstructed view of an intersection, and safe gaps to join or cross traffic streams resulted in safer driving performance. Additionally, one-way street message warnings resulted in fewer route errors. The results tended to be generally the same for all age groups. Further, the presence of the system did not reduce driver workload. They concluded that the system, while promising, needs longer evaluation periods to determine its long-term effects.

Rauch, Gradenegger, and Krüger (2008) analyzed the efficiency of different strategies for the interaction with in-vehicle devices. Using a motion-based driving simulator, 24 drivers completed a test course. A control group of eight participants provided baseline driving condition data (no additional task demand), while the other group, a dual-task group, performed a secondary navigation task (e.g., a hierarchical menu navigation task simulating interaction with a typical in- vehicle information system) while driving. Variables included the criticality of the driving situation, road type, and predictability of the critical situation (easy, moderate, hard to predict). At predetermined points, drivers were offered the choice to perform the task. The offer was given just before a critical situation (e.g., just after a parked car started to indicate but was still standing) or in a non-critical situation (e.g., on-road segments between critical situations). Drivers were permitted to determine whether or not the situation was suitable for executing a task and also when to interrupt it. The secondary task offer was signaled by a question mark shown in the HUD on the front scene. Researchers allowed drivers 3 s to decide whether the situation was suitable or not for the secondary task, according to the situational demands. The secondary task was started when the driver pulled to the right a joystick on the vehicle's middle console, at which time a "start display" message was presented on the visual display located at the lower position on the vehicle's middle console. No action was required to reject the task. Researchers found that drivers were able to adapt their secondary task behavior to situational demands. Participants anticipated potential conflicts that resulted in secondary task rejection or delay in critical situations. These strategies were found to be successful for maintaining driving safety. Rauch et al (2008) concluded that situational assessment prior to the start of a secondary task and that adequate monitoring of situational developments during task execution are relevant processes for situational awareness in this context.

Reimer, Mehler, and Coughlin (2010) used heart rate as an objective physiological arousal measure along with self-reported ratings to evaluate the extent to which two vehicle-parking-assist technologies impacted driver stress. The first technology was a semiautonomous system for parallel parking that detected appropriately sized parking spaces and actively steered the vehicle into the parking space while the driver controlled the throttle and brake. The second technology was a cross-traffic warning system that was designed to alert drivers of encroaching vehicles when backing out of parking spaces. Each technology was tested using 42 participants consisting of three gender-balanced age groups (20–29, 40–49, and 60–69). Participants were asked to park with and without the technology.

Across gender and age groups, participants reported lower stress levels and exhibited lower average heart rates when using the assistive parallel parking technology. When using the cross-traffic warning system, findings suggested some reduction in stress levels, although not at a statistically significant level. However, when drivers used the cross-traffic alert system, the likelihood of their stopping and yielding to an approaching vehicle was increased, which the researchers noted, is a potential safety benefit in that this behavior could potentially reduce the likelihood of accidents.

Merat, Jamson, Lai, and Carsten (2012) conducted a driving simulator study to compare the effect of changes in workload on performance in manual and highly automated driving and also to determine changes in driving state exhibited as variations in blink patterns. In this study, 50 participants operated a motion-based driving simulator in manual and highly automated driving modes. Automation included longitudinal control through ACC with a fixed target speed of approximately 113 km/h (70 mi/h) and an adjustable headway that defaulted at 1.5 s. The lateral controller design was similar to a lane-keeping assistance system which kept the vehicle centered in the lane. Participants engaged the automation via a button on the steering wheel and could disengage the automation by pushing the button, turning the steering wheel more than 3 degrees, or pressing the brake pedal. Workload was manipulated using both non-driving cognitive tasks and by forcing a lane-change event at designated points. The effects of automation and workload on drivers' awareness of driving environment were further assessed through a comparison of participant responses to the critical incidents in manual and automated driving. The results demonstrate that, in the absence of a secondary task, drivers' responses to critical incidents were similar in both manual and highly automated driving conditions. Therefore, they concluded that when attention was not diverted to distracting secondary tasks, driving performance was not adversely affected by highly automated driving. Similarly, in the absence of a secondary task, the rate and duration of blinks for automated driving were similar to those for manual driving. However, drivers displayed the lowest performance when they were required to regain control of driving in the automated mode while distracted by the secondary task. Blink suppression was highest in automated driving with both secondary tasks and the critical event. The authors posited that the findings suggest that the change in demand from the driving task and an unexpected need to regain vehicle control provoked drivers to obtain as much information as possible about the visual scene, where moments before their attention may have been directed mostly toward the secondary task. Researchers concluded that although further studies using a combination of subjective and objective measures are required to understand changes in workload experienced by participants in highly automated driving conditions, using nonintrusive tools for the real-time observation of driver workload is an advantage in warning drivers of dangerous overload or underload situations.

Kaber, Liang, Zhang, Rogers, and Gangakhedkar (2012) assessed the effects of visual, cognitive, and simultaneous visual and cognitive distractions on operational (braking, accelerating) and tactical (maneuvering) control of vehicles. Using a driving simulator with an integrated head-mounted eye/head tracker, 20 participants drove in lead-car following or passing scenarios under four distraction conditions: without distraction, with visual distraction, with cognitive distraction, and with simultaneous visual and cognitive distraction. Visual distraction was found to increase driver workload through more complex gaze behavior; however, drivers appeared to compensate for visual distraction by increasing headway times. Cognitive distraction also increased workload by dividing driver concentration between roadway and secondary tasks, but did not result in steering errors as high as those found with visual distraction. Tactical control behavior required greater workload than operational control, was more sensitive to distraction, and less conducive to adaptation. However, drivers engaged in tactical control were able to perceive the higher workload when presented with simultaneous distractions and, as a result, prioritized the primary driving task, which then resulted in slower responses to secondary tasks. The results indicated that drivers experienced higher perceived workloads for passing tasks than for following tasks, with drivers having slower responses to secondary distraction tasks as workload increased during passing tasks. Additionally, the simultaneous distraction task resulted in the greatest demand and steering errors in both driving tasks. Visual distraction resulted in more off-road glances, higher workloads, and longer headway times. Further, while cognitive distraction was found to increase driver workload, researchers did not find steering errors as high as those found with visual distraction. Kaber et al. (2012) concluded that tactical control of a vehicle results in greater workload than does operational control.

The lane change task (LCT) is a commonly used technique in studies of manual driving, especially under secondary task conditions (ISO 26022, 2010). The LCT allows researchers to examine quantitatively the effect a secondary task has on primary driving-task performance degradation. Benedetto et al. (2011) analyzed psychophysiological variables that may indicate visual workload. Specifically, they studied the effects of IVIS usage on eye blinks in a simulated LCT. Participants were asked to perform the LCT in single- and dual-task conditions. The dual-task condition involved interacting with the IVIS while performing the LCT. The specific secondary task chosen was the Surrogate Reference Task (a search task presented on an in-vehicle monitor; Mattes, 2003). Results indicated the blink duration to show a Gaussian-like distribution in single-task conditions, while the distribution shifted to the left of the curve in dual-task conditions. They concluded that blink length inhibition may have occurred to avoid visual information loss. Additionally, blink rates were shown to reflect visual workload and time on task: shorter blink rates were associated with IVIS interaction during driving while long blink rates occurred with greater frequency as the time spent driving increased.

Adapting the LCT to automated driving, Spießl (2011) and Spießl and Hussmann (2011) developed the Autonomous Lane Change Test (ALCT) to measure error recognition performance during an automated drive in a driving simulation environment using a set of objective metrics including mean response time, missed errors, false interventions, and driver activity load index (DALI). During the ALCT, the user is asked to perform different secondary tasks while the vehicle performs lane changes in an automatic fashion (although with an error rate of 10%). In developing the ALCT, Spießl (2011) conducted a user study and found that an analogue HUD following the paradigm of Augmented Reality was best for visualizing the vehicle's trajectory. Spießl and Hussmann (2011) conducted two studies using the ALCT to explore drivers' reactions to automation errors (i.e., in a lane change situation versus a lane-keeping situation) in a simulated driving scenario when also engaged in a secondary task (i.e., entering a task into a navigational system; selecting a target directly on an interactive map; making a phone call; listening to an audio book; reading a long, unstructured text on a display; reading a single line of text). In the case of automation errors, two different interaction concepts were tested:

1. *LO version* in which the steering wheel served as a binary input device. The steering wheel remained in a neutral position during the course of the study and all lane changes. When an error occurred, drivers were required to turn the wheel at least 90 degrees in the correct direction and back to its original position.
2. *HI version* in which the steering wheel actually controlled the car on the track and the steering action was automated; the car steering wheel moved according to the path of the virtual car, as it would in an actual car with automated lateral control. The direct haptic feedback provided participants with information about the vehicle's trajectory. When an error occurred, participants could override the automation by turning the steering wheel to the desired direction with a force greater than 3 N·m for 0.5 s.

Spießl and Hussmann found that participants used less time to redirect their attention towards the road after an error when driving with haptic steering wheel feedback. Additionally, participants remained more engaged in the driving task with the haptic feedback. They concluded that both tested scenarios provide valid methods for examining the influence of secondary tasks in an automated driving situation; however, the haptic interface version with actuated steering wheel was recommended.

Driver monitoring can be used to supplement or augment the performance of the vehicle DVI by allowing it to adapt to current conditions. Fletcher and Zelinsky (2009) proposed and validated an automated driver-assistance system that is not only responsive to the driver's actions but also designed to correlate the driver's eye gaze with road events to determine the driver's observations. This driver observation monitoring system enabled an immediate in-vehicle system to detect and act on driver inattentiveness. The benefit of such a system was that it suppressed redundant warnings and canceled warnings using eye glances. The end result was an automated co-driver system capable of detecting missed road events and warning the driver appropriately.

Lu et al. (2013) conducted three meta-analyses to contrast performance on an ongoing visual task and interrupting tasks as a function of task modality (i.e., auditory vs. tactile, auditory vs. visual, and single modality vs. redundant auditory-visual). The purpose of this study was to integrate empirical data showing the effects of interrupting task modality on the performance of an ongoing visual-manual task and the interrupting task itself. They noted that operators in a variety of domains are often required to monitor the performance of a number of automated systems, which often results in data overload in the visual channel. At the same time, operators also need to cope with a number of tasks and responsibilities which are accompanied by an increased risk of interruption of ongoing tasks and associated performance tasks. The use of a multimodal interface design that distributes information across vision, audition, and touch may address the challenge of data overload and the need for effective interruption management (Lu et al., 2013; citing Oviatt, 2003; Sarter, 2002). Lu et al. examined 68 studies and considered six moderator variables (i.e., variables which affect the relationship between two other variables; in this case, interrupting task modality and performance on the ongoing and interrupting tasks):

1. Ongoing task workload (high vs. low);
2. Interrupting task design complexity (level of uncertainty within the signal, i.e., low-complexity informing of an event vs. high-complexity requiring an action);
3. Interrupting task urgency (alarm vs. notification);
4. Interrupting task code (spatial, i.e., "spatial relationships between stimulus components such as left-right" vs. categorical, i.e., "the extracted information [that] has symbolic meaning or refers to identity within a category");
5. Auditory permanence (permanent; e.g., a repeated tone vs. a transient tone); and
6. Visual angle of separation (the angle of separation measured by the number of degrees between the ongoing task's center of focus and the interrupting task's visual display).

Results indicate that response times were faster for tactile interrupting tasks in the case of low- urgency messages. Also, accuracy was higher with tactile interrupting tasks for low-complexity signals but higher with auditory interrupting tasks for high-complexity signals. Further, redundant auditory- visual combinations were preferable for communication tasks during high workloads and with a small visual angle of separation. Based on these findings, the authors concluded that their findings highlight the importance of these moderator variables in predicting the effects of interruption task modality on ongoing and interrupting task performance. Additionally, these findings can be used to inform the design of multimodal interfaces in data-rich, event-driven domains.

Lu et al.'s (2013) findings are supported by an earlier effort. Lu, Wickens, Sarter, and Sebok (2011) completed a meta-analysis of 23 studies to determine the ability of the interrupting task/signal (IT) to capture attention without compromising performance of the ongoing task (OT). Further, they examined how these goals are moderated by the modality of the IT (auditory versus tactile), and how the modality difference may be modulated by other factors, such as OT workload (high versus low); IT complexity (number of varied parameters within the signal); IT decision complexity (level of uncertainty within the signal, i.e., does it cue to an upcoming event or provide information about which of two or more possible events will occur); IT urgency (alarm versus notification); IT processing code (spatial or categorical); and tactile stimulus location. The comparison across studies was completed using a complementary metric representing the ratio between tactile and auditory IT performance, and where possible, OT performance between both modalities. The researchers found that, averaged across all conditions, tactile interruptions are responded to 6% faster than auditory interruptions. Additionally, no statistically significant difference exists in accuracy between auditory and tactile task performance as does no speed-accuracy trade-off. Recommendations regarding the moderator variables include using tactile cues for low-complexity IT conditions and for notification alerts, while auditory cues are recommended for high-complexity IT conditions and for urgent alerts.

Abbink, Mulder, and Boer (2012) argue that force-feedback (haptic) shared control as a DVI can intuitively provide for shared control between human-automation interactions. They noted that research shows that haptic shared control can lead to short-term performance benefits (e.g., faster and more accurate vehicle control; lower levels of control effort; reduced demand for visual attention). However, while continuous intuitive physical interaction inherent in haptic shared control is expected to reduce long-term issues with human-automation interaction, additional research is needed to explore this theory. Potential areas presented for future research associated with the long-term use of haptic shared control systems include trust, overreliance, dependency on the system, and retention of skills.

Additional efforts by Mulder, Abbink, and Boer (2012) included an investigation of force-feedback (haptic) shared control as a DVI that can intuitively share control between drivers and an automatic controller for curve negotiation. They noted that the conventional binary switches between supervisory and manual control have many known issues and that haptic shared control is a promising alternative. They argued that manual control and automation should be connected in one coherent framework grounded in haptic shared control. To explore this, a driving simulator experiment was conducted to compare participants' curve negotiation behavior during shared control to curve negotiation during manual control, as well as to three variations of haptic force feedback steering of an automatic controller. The controller provided force feedback, which increases steering forces based on a single- point look-ahead controller, and stiffness feedback, which increases stiffness around the target steering angle. Within this system, the stronger the stiffness, the less the driver can do and the more automated the system acts. Thus, at a high level of haptic authority, the steering torque would be strong enough to navigate the vehicle automatically along the controller's reference trajectory. The controller used within the experiment to generate haptic shared control was identical to the controller previously used by the researchers (Mulder et al., 2008; Mulder & Abbink, 2010). Results indicated that the main beneficial effect of haptic shared control compared to manual control was that less control activity was needed for realizing an improved safety performance. While the automated condition improved safety performance, because the need for human control activity was removed, drivers were put in a supervisory position. Mulder et al. (2012) concluded that haptic shared control kept the driver in-the-loop, with enhanced performance at reduced control activity, which mitigated the issues associated with full automation. Additionally, Tsoi, Mulder, and Abbink (2010) presented a shared control lane-keeping haptic guidance system which was capable of continuously and smoothly supporting lane changes. A fixed-base driving simulator experiment was conducted to assess the effects of the support system on lane-change behavior when haptic guidance was provided during lane-keeping and lane-changing situations. Tsoi et al. found that the system provided drivers with the benefits of haptic guidance for lane keeping while also allowing drivers to smoothly change lanes.

Further curve-related efforts have explored the benefits and limitations of car-following support systems (Abbink et al., 2008) as well as the use of haptic controls by young, inexperienced drivers (Mulder, Abbink, & Boer, 2008) and elderly drivers (Mulder & Abbink, 2010). When looking at the effects of haptic guidance on curve negotiation behavior of young, inexperienced drivers, Mulder, Abbink, and Boer (2008) developed a haptic guidance system that continuously generated low forces on the steering wheel which required active input from the driver to safely negotiate curves. They tested this system in a fixed-base driving simulator and found that the haptic guidance resulted in reduced and smoother driver steering behavior (Mulder et al., 2008). With elderly drivers, Mulder and Abbink (2010) found that although the haptic guidance system for curve negotiation resulted in a small increase in curve negotiation performance with less control activity, the haptic guidance system resulted in a relatively large increase in steering forces, which could potentially be a disadvantage for elderly drivers.

Building upon these findings, Mulder and Abbink (2011) investigated how drivers respond to faulty control of the previously studied haptic shared system (i.e., Mulder et al., 2008; Mulder & Abbink, 2010) when trying to negotiate the vehicle around an obstacle suddenly appearing on the road. Using a fixed-base simulator, drivers were asked to negotiate a winding road that had obstacles at either end. Drivers were required to respond in instances when the support system responded too late and to override the faulty late response of the system. Comparing driver responses during the faulty control with responses when drivers did not share control with the haptic system, they found that for a time-to-contact of 1.4 s, haptic shared control reduced the hit rate with obstacles from 21.2% to 15.2%. Further, under the faulty conditions, the haptic shared control allowed the driver to understand the fault in the system and to respond in 35% of the cases.

Penna, van Passen, Abbink, and Mulder (2010) also presented findings associated with the use of a haptic interface during collision avoidance maneuvers. Penna et al. (2010) conducted an experiment using a fixed-base simulator wherein drivers had to avoid an obstacle through either a right or left maneuver. The collision-avoidance system supported either solution through a haptic steering wheel interface featuring torque feedback and stiffness feedback. The researchers found that the haptic feedback effectively reduced the number of crashes and also decreased response time, control effort, and activity in the most critical situations.

Brandt, Sattel, and Böhm (2007) developed a proof-of-concept steering wheel providing force- feedback with potential field path planning and path following for lane-keeping and collision- avoidance assistance. Within the proposed system, drivers could influence the path planning through the haptic interaction with the vehicle. The proposed lane-keeping system and collision- avoidance systems required driver control over longitudinal vehicle guidance while the driver and assistance system work in tandem for the lateral vehicle guidance. The proposed path-planning algorithm, the corresponding path-following controller, and the assistant torque characteristics were tested using a fixed-base simulator. Drivers were required to complete a single drive of approximately 4 min along a curvy road course containing four non-moving obstacles which appeared at different locations and different distances in front of the host vehicle while performing a secondary task (a visual search task on an in-cabin monitor). The results indicated good driver acceptance for the lane-keeping assistance concept; however, it was noted that the proposed collision-avoidance assistance system needed improvement.

Griffiths and Gillespie (2005) demonstrated that adding automation through haptic display could improve performance on a primary task and reduce perceptual demands or free attention for a secondary task. Three experiments were conducted using a force-feedback steering wheel on a fixed- base driving simulator. This level of automation was compared to a copilot that assisted with lane following by applying torques to the steering wheel. The results of the lane-following task experiments indicated that haptic assist improved lane following by at least 30% while reducing visual demand by 29% and improving reaction time in a secondary tone localization task by 18 ms. The authors concluded that this research may influence the design of DVI based on haptics that support human/automation control sharing better than traditional push-button automation interfaces.

Lervag, Moen, and Jenssen (2010) studied DVI for lane-keeping systems, specifically DVI concepts for driver assistance systems that alert the driver when the vehicle is crossing lane boundaries. They conducted a simulator study to explore two alternative DVI solutions based on tactile feedback, with vibration motors placed in the driver's seat and steering wheel. The study found that, although subjects experienced the systems for lane support as useful and effective in terms of driver behavior and traffic safety, subjects expressed low willingness to pay to install such a system in their personal vehicles. However, they supported the mandatory installation of LDWS in new cars. Additionally, the authors found that subjects were most receptive to the tactile steering wheel followed by the tactile seat and were least receptive of the standard audio/visual system. Further, they found that subjects with lower sensation-seeking tendencies had more trust in the system and were more positive to the LDWS than were subjects with high sensation-seeking tendencies.

Radke et al. (2013) analyzed human dynamics perception in automated longitudinally controlled passenger vehicles. They noted that the derivation of a continuous dynamics measure has been suggested to be a key to the implementation of a well-designed DVI. A two-step empirical approach was adopted. The first step involved the evaluation of longitudinal vehicle dynamics while the second step served as the basis of an adjustable longitudinal control DVI. The authors conducted real-world tests along a 15.8-km section of road in Germany. A preliminary study was designed to identify vehicle dynamics parameters characterizing styles of driving. Six participants drove the test vehicle manually. Participants were instructed to drive in *explicitly undynamic* (i.e., strongly comfort-oriented and homogeneous) and *explicitly dynamic* (i.e., very sportive and ambitious) styles. Relevant parameters (resulting from the analyses of the driving styles and recorded controller area network, or CAN, values) were merged into a single value called the Dynamics Factor which was used for the objective measurement of human subjective dynamics perception in automated longitudinally controlled passenger vehicles. The second phase involved 31 participants, who were responsible for lateral vehicle control while longitudinal control was autonomously performed by the vehicle. Manual brake and accelerator pedal use was limited to potential emergency situations. The dependent variable was the subjective perception of the amount of dynamics, which were quantified by participants after each track segment via a continuous percentage value. The results indicated that the Dynamics Factor explained 70.5% of the participants' subjective dynamics variance while driving in an automated longitudinally controlled passenger vehicle on rural road tracks. Further, they found that parameters such as sight distance and route knowledge have a significant influence on drivers' subjective dynamics perception. Also, the correlation of the Dynamics Factor with the participants' subjective ratings proved the existence of the co-driver effect (i.e., an effect resulting from systematically increasing the amount of subjective dynamics perceived when displacing the driver from an active into a passive role). It was noted that the proposed DVI was designed to adjust the vehicle's longitudinal dynamics behavior to the driver's expectations and preferences. In doing so, the DVI keeps the driver actively involved in the control loop and in charge of the driving task.

As described in the preceding works, both L2 and L3 operations present the potential for greater performance in handling small driving errors that human drivers typically commit. These errors are committed for a number of reasons, such as distraction, confusion, disengagement, or fatigue, and are generally below the level of safety concern. An example of a small error is a steering reversal that does not result in a lane exit. While the human may be subject to many factors that affect performance, it is assumed that any L2 or L3 production automated system will be able to provide a constant level of environmental monitoring and always will be able to provide an appropriate response provided that the lane sensing is accurate.

Trust in Automated Vehicles

Automation, across any domain, presents difficulties when individual users' reliance on the automation is not properly calibrated to the performance. The relationship between trust and automation can lead to several different outcomes, as defined by Lee and See (2004). *Calibrated trust* describes a system in which the user's trust matches the automation capabilities. Calibrated trust supports appropriate application of the automation. *Overtrust* describes a system in which the user's trust in the automation exceeds the actual capabilities. Overtrust can lead to misuse of the automated system, where the driver applies the automation to a roadway environment that is outside of the automation's operational scenarios. *Distrust* describes a system in which the user believes that the automation's performance is less than it actually is. Distrust can lead to disuse of the automation, removing the possible benefits of the automation.

Although trust in automation has been studied across a number of domains, the applicability of these findings to driving is questionable due to fundamental differences between the operational characteristics of the different automated systems. Perhaps the best analogy may be drawn from aviation. Modern aircraft have multiple highly automated systems onboard. Modern autopilot systems have the ability to automate almost all phases of flight. This automation allows the pilot to transfer to a monitoring role. However, the analogy begins to fail when considering the fact that airspace is somewhat controlled, routes are mapped, and the potential for other airspace users to perform unexpected threatening maneuvers in very close ranges (i.e., under 2 s.) is not typically present. Therefore, while the lessons learned from other automation domains should be considered in approaching L2/L3 vehicle automation, they may not necessarily be directly applicable. More research on the topic of drivers' trust in L2/L3 automation could provide benefits towards proper use of in-vehicle automation. Ultimately, the issue of drivers' trust in L2 and L3 automated vehicles is a significant factor in understanding how these technologies are both used, and misused, in the real world.

Studies Associated with System Trust and Acceptance

Several studies have examined drivers' trust and acceptance of automated systems. These studies have addressed how issues such as shared goals, information presentation, and imperfect automation all interact with trust and acceptance of automated systems. In addition, issues of communicating uncertainty in automated systems are discussed. The following presents a more detailed account of these findings.

Verbene et al. (2012) examined whether describing ACC as sharing driving goals affected the perceived trustworthiness and acceptability of ACC and whether trustworthiness and acceptability of ACC depends on its automation level. Participants were presented with perceptions of ACC automation levels (as opposed to functional differences) using a computer-based presentation with participant feedback obtained throughout. Results demonstrated that ACC systems that share the driving goals of participants are viewed as more trustworthy and acceptable than are ACC systems that do not share the driving goals of participants. Additionally, the automation level of ACC affects both the trustworthiness and the acceptability of ACC. Those systems that both take control over vehicle actions and provide information are judged as more trustworthy and acceptable than are systems that take control over vehicle actions without providing information. Finally, the researchers found that results indicated that trustworthiness is a mediator for both the effect of shared driving goals and the effect of automation level on the acceptability of ACC. This study is significant in that it provided a first indication that both sharing goals and providing information increases the trustworthiness and acceptability of smart systems.

Beller et al. (2013) evaluated whether communicating automation uncertainty improved driver- automation interaction. The researchers noted that few studies have evaluated the presentation of uncertainty in a non-decision-making task (citing McGuirl and Sarter' s (2006) automation study which utilized a line graph to display system confidence and Seppelt and Lee's (2007) investigation of uncertainty representations in the driving context). To determine whether uncertainty information would improve the appropriateness of human behavior in cooperation with highly automated vehicles, Beller et al. conducted a driving simulator experiment. During the experiment, participants interacted with a highly automated driving system (i.e., the system offered both longitudinal and lateral control in the form of ACC and run-off-road prevention) while driving through a fogged, two-lane highway scenario. Additionally, participants also engaged in secondary tasks (i.e., a visual search task) while required to cooperate with the automation to drive safely. During the 60-min driving scenario, participants were subjected to four conditions with varied degrees of system reliability and certainty. When the automation was unreliable, the system sometimes produced either false or no-braking responses that had to be overruled. When conditions were unclear, the system's uncertainty symbol (a face with an uncertain expression and hand gestures) was presented. After system encounters, participants were asked questions to determine trust in automation and system acceptance. The presentation of uncertainty information increased the time to collision in the case of automation failure. Additionally, the data indicated improved situation awareness and better knowledge of fallibility for the experimental group. As a result, automation with the uncertainty symbol received higher trust ratings and increased acceptance. The authors concluded that the presentation of automation uncertainty through a symbol improved overall driver-automation cooperation. These findings suggested that driving automation systems could benefit from the display of reliability information.

Weinstock, Oron-Gilad, and Parmet (2012) incorporated aesthetics manipulations in an imperfect in- vehicle automation system in order to explore how aesthetics can decrease negative effects of errors on trust, satisfaction, annoyance, and human-automation cooperation perceptions. Using a PC-based experimental system simulating a navigation system, 141 participants were asked to navigate through four conditions (100% or 85% accuracy, and an aesthetic or non-aesthetic system). Results indicated that participants had decreased perceptions of trust, satisfaction, and human-automation cooperation in the system with 85% accurate performance. The annoyance rating was only identified in the aesthetic system regardless of accuracy. They concluded that, compared to the impact on trust and satisfaction that aesthetics plays in mobile commerce or websites, aesthetics has a limited impact on the trust and acceptance of automated systems; however, additional research is needed to determine the extent of the impact that aesthetics has on trust in imperfect automated systems.

Blair, Sandry, and Rice's (2012) system-wide assessment theory extended system-wide trust theory to make predictions regarding in-vehicle automated devices. They tested for a system-wide effect for in- vehicle automation by asking participants to provide trust ratings as well as perceived reliability ratings for multiple in-vehicle devices. Two surveys were used to assess the impacts of imperfect automation on trust and perceived reliability. The first survey asked participants to indicate their levels of trust and perceived reliability in regard to an in-vehicle automated device that was non-specific. Additionally, participants were asked for their perceptions about four additional devices for which they were given detailed information (i.e., a new blind-spot camera in the side mirror, a new GPS system programmed to know the speed limit and give a warning if the driver is speeding, a warning device to lets the driver know if he or she is driving too close to the car in front, and a driver's seat that automatically adjusts to the driver's body size). At a later time, participants were asked to complete a second survey which presented specific in-vehicle automated devices and additional devices. Findings indicated that participants' initial perceptions of reliability levels impacted both subsequent reliability estimates and subsequent trust ratings of both specific and non-specific in-vehicle automated devices. The authors concluded that manufacturers need to keep in mind the implications associated with having an unreliable device in the same vehicle as a reliable device; users may be dissuaded from using a safety-related device based on perceptions and trust levels associated with a previous device.

Using a driving simulator, Lees and Lee (2007) examined the influence of collision warning systems and distraction on driver performance during non-critical and critical events. They noted that false alarms and unnecessary alarms (also termed nuisance alarms or alarms that occur in situations judged hazardous by the algorithm but not by the driver) may limit the effectiveness of the collision warning system. Participants were subjected to four types of situations: alarm, false alarms, unnecessary alarms, and critical medium-onset alarms (triggering all collision warning systems). Findings indicated that context influenced compliance with alarms, with drivers complying more with unnecessary alarms that occurred in the context of a non-critical driving event, compared to false alarms, which occurred in seemingly random fashion. The results suggest that the false alarms diminished trust and compliance, whereas the context associated with unnecessary alarms fostered trust and compliance during subsequent events. Further, they noted that these findings suggest that current warning descriptions based on signal detection theory need to be expanded to represent how different types of alarms affect drivers.

Seppelt and Lee (2007) suggest that providing drivers with continuous information about the state of automation may be more effective than warning drivers of imminent crash risk when the system fails. Using a driving simulator, they evaluated the effect of automation (manual, ACC control) and display (a display designed using EID, or no display) on ACC reliance, brake response, and driver intervention strategies. The results demonstrate that drivers in traffic conditions relied more appropriately on ACC when the EID was present, and the EID display promoted faster and more consistent braking responses which resulted in safe following distances and no collisions. In the manual scenario, the EID display proved useful to drivers in terms of time headway maintenance which reduced driving demand and promoted more consistent and less variable car-following performance. In a later study, Seppelt (2009) built upon these findings through the development of a concurrent information display that provides purpose, process, and performance information to drivers. Again, the goal of the system was to keep drivers continuously informed as opposed to alerting them with occasional alarms or warnings.

Larsson (2010) conducted a mainly qualitative survey of ACC users to identify how and when control is transferred between the driver and the ACC in naturalistic conditions, the extent of experienced drivers' understanding of the ACC system limitations, and whether or not the drivers ever forgot if the system was activated or not. The system in question was functional at speeds over 30 km/h and could be disengaged by pushing the brake pedal or by driving slower than 30 km/h. The set speed was presented in the instrument panel. Additionally, when losing radar contact, an icon indicating a radar lock on the vehicle in front disappears, and the car starts accelerating to the set speed. Surveys were sent to manufacturer-identified owners of a particular car model in the premium segment. Of the 130 respondents, 9 declared that they never used the ACC system. Larsson found that drivers used the system in the same situations as they would ordinary cruise control and that many disengage the system in dense traffic. Although mentioned in the user manual, 36 respondents were unaware of any system limitations. Those noting limitations included ones not mentioned by the manual, including difficulties in roundabouts, coming to the top of a hill, being overtaken by another car, if the vehicle in front brakes sharply, and large vehicles in the side lane. Additionally, 31 respondents reported having forgotten whether or not the system was active. Other respondents voiced concerns over the system's operation (e.g., do the brake lights come on). Larsson concluded additional education may be necessary and provided suggestions for future research focusing on who is responsible for making the driver aware of the system's operational range and functionality.

Larsson (2012) further examined these issues through a survey of 130 ACC users and found that the longer drivers used their systems, the more aware of its limitations they became. Moreover, the drivers reported that ACC forced them to take control intermittently. Larsson posits that current ACC systems may be less detrimental to driver performance than previous research has suggested because the less-than-perfect system forces the driver to reclaim control from time to time, which may help the driver to stay in the loop.

Through a series of studies, the California Partners for Advanced Transportation TecHnology (PATH) program has explored drivers' preferred following distances with regards to CACC technology. Building upon the previously developed methods (Shladover et al., 2009), Nowakowski, O'Connell, Shladover, and Cody (2010) conducted a field test to determine whether or not drivers would be comfortable with following time gaps under 1.0 s provided by a CACC system. They found that while drivers were generally comfortable with and selected the sub-second time gaps offered by the CACC system, significant differences existed between the preferences of males and females. The study found that the shortest CACC time-gap setting (0.6 s) was used over 55% of the time, with most males preferring this time-gap setting while most females preferred the longer (0.7 s) time-gap setting. Additionally, participants indicated that the time-gap settings offered by ACC systems resembled time gaps they would keep while driving manually in light to medium traffic, while CACC systems were closer to the time gaps that they would keep while driving manually in heavy traffic, a statement confirmed through an examination of the baseline (manual) driving. Driver preferences regarding shorter gap time settings enabled by CACC technology are further discussed in Nowakowski et al. (2011).

Adaptive Cruise Control and Forward Collision Warning Systems

Prior Monte Carlo simulations have estimated that a nationwide deployment of FCW systems in heavy vehicles could reduce the number of rear-end crashes by approximately 21% (Fitch et al., 2008). A benefit-cost analysis that evaluated the benefits in terms of crash cost avoidance and costs in terms of technology costs indicated that for every dollar spent carriers would get more than a dollar back in benefits ($1.33 to $7.22, depending on vehicle miles traveled, estimates of system efficacy rates (i.e., crash prevention rates), and technology purchase prices) (Murray, Shackelford, & Houser, 2009). While these types of examinations have focused on the safety outcomes of the technologies used, they have not addressed the underlying driver performance issues that arise from FCWs. The works presented in this section address advanced technologies such as ACC, FCW, and LDWS from the perspective of driver behavior and performance.

The NHTSA-sponsored Road-Departure Crash Warning System (RDCW) FOT (Wilson, Stearns, Koopmann, & Yang, 2007) set out to investigate if the RDCW – which warned drivers when they were either drifting out of their lane or about to enter a curve at an unsafe speed – could help reduce road- departure crashes. A lateral drift warning (LDW) subsystem alerted participant drivers when their vehicle was in danger of departing the roadway, and a curve-speed-warning (CSW) subsystem issued an alert when the vehicle was in danger of losing control in an upcoming curve. Using a video camera to estimate distances, the LDW monitored the subject vehicle's lane position, lateral speed, and available maneuvering room. In addition, using side and forward radar (two units each), the LDW detected adjacent and upcoming objects. The CSW used GPS technology to determine the subject vehicle's location, and a road database determined the curvature of the road several seconds in front of the vehicle. The CSW also predicted the most likely path that the vehicle would travel. The study vehicles were outfitted with two unobtrusive cameras, extra sensors, and data acquisitions systems, which saved 8 s of buffered video when an alert occurred. The RDCW FOT collected 130,000 km of driving data from 78 male and female participants ranging in age from 20 to 70. Participants used the FOT vehicles in place of their own vehicles for 25 days. A baseline period of 6 days (in which no alerts were issued to the drivers) occurred at the beginning of the study followed by a 19-day treatment period with alerts provided. FOT participants also completed surveys before and after their FOT experience, discussed their opinions of the RDCW, and 32 of the 78 participants were involved in four 2-hour focus groups.

The performance and safety benefits of the system, as well as driver acceptance, were examined. Lighting and precipitation influenced the availability of the LDW system. While the system was available 76% of the time (i.e., 76 out of every 100 miles) on freeways, it was only available 36% of the time on non-freeways. Likewise, it was available 56% of the time during dry, daytime conditions, but only 4% of the time during wet, nighttime conditions. However, CSW availability was high: with 99% availability on freeways and 94% on non-freeways. Both the LDW and the CSW had inaccuracies in their alerts; e.g., one in three LDW alerts was a false positive, and the CSW system failed to alert one out of four cases when a vehicle approached a curve with excessive speed. However, participants gave the following features favorable ratings: LDW and CSW alert timing, LDW and CSW missed alert frequency, and LDW false-positive alert frequency. The FOT data indicated that with the RDCW activated a 10- to 60-percent reduction in departure conflict frequency was observed at speeds over 88 km/h (approximately 55 mi/h). Assuming 100-percent device availability and deployment, an annual reduction of 9,400 to 74,900 road-departure crashes was forecast.

The 5-year Automotive Collision Avoidance System (ACAS) FOT (GM, 2005) looked at the potential implications of FCW and ACC systems from both a traffic perspective and a driver acceptance perspective. The ACAS developed in this project consisted of FCW and ACC subsystems, both of which used forward-looking radar (FLR). The DVI included a HUD to show ACC- and FCW-related information and a data fusion system designed to accurately determine forward road geometry. The 96 drivers who participated received training on the ACAS system and then drove the deployment vehicles as their personal vehicles for a 3- or 4-week period. During the first week, the ACAS features were not available; after the first week, these features became available for use.

Early in the study driver dissatisfaction resulted from the false alarms (termed nuisance alerts), many of which were due to stationary objects along the roadway that were incorrectly classified as threats. Subsequent revisions of the software algorithms addressed this concern.

The data collected during this study included 1.4 terabytes of information, as well as subjective evaluations resulting from participant interviews, questionnaires, and focus groups. Results showed that the FCW and ACC subsystems reduced the incidence of tailgating as compared to what is seen in manual driving, although this effect was restricted to daytime and freeway driving. However, evidence that the FCW and ACC systems reduced approach conflict behavior was mixed. Driver acceptance of the ACC systems was uniformly high, but mixed for the FCW, possibly owing to the workload and stress reduction afforded by the ACC, and credibility issues resulting from the false alarms generated by the FCW system. Study authors suggested that a larger FOT with a longer-term exposure to an ACAS system is needed.

The Crash Avoidance Metrics Partnership (CAMP) FCW efforts provided further insight into the development of a DVI to accompany an FCW, how driver behavior and judgments compare under on- road versus simulated approach conditions, and recommendations for simulated scenario design. Keifer, Cassar, Flannagan, Jerome, and Palmer reported on CAMP FCW tasks two and three, noting that this research had multiple objectives, including (2005, p. 1):

- To examine the extent to which alert effectiveness is influenced by a wide range of factors such as driver characteristics, environmental factors, interface design, distraction activity, kinematic conditions, and training/false alarms.
- To use visual occlusion techniques under real approach conditions to further understand the driver's decision-making and avoidance maneuver behavior in rear-end crash scenarios.
- To provide a calibration dataset for understanding how driver behavior and judgments compare under on-road versus simulated approach conditions.

In an effort to simulate realistic driving conditions, the authors used a surrogate target, test track methodology with surprise braking trials. Because simulated approach conditions involving degraded visual scene properties had been shown to influence time-to-collision (TTC) judgments and, subsequently, driver's perceptions of crash threats, the authors used a TTC judgment occlusion technique (the driver's vision was occluded during the last phase of an in-lane approach to a lead vehicle, after which drivers were to press a button the instant they felt they would have collided with the vehicle ahead) and a first-look occlusion technique (the driver's vision was occluded during the initial phase of an in-lane approach to a lead vehicle, after which the driver's vision was suddenly opened and the driver was instructed to avoid colliding with the lead vehicle). The authors drew the following conclusions (Keifer et al., 2005, pp. 66–67):

- Based on test driver intervention rates during surprise trials, the alert timing approach evaluated, coupled with a single-stage, dual-modality (auditory plus visual) FCW alert, was found to be robust, effective, and judged appropriate across the wide range of conditions evaluated.
- Results from the time-to-collision (TTC) and first look visual occlusion studies suggested that, provided the driver is looking toward the lead vehicle, the driver can quickly assess TTC and make the appropriate crash avoidance maneuver under the alert timing assumptions evaluated.
- The first look method appears to be a valid, efficient, and promising method for exploring the consequences of later FCW alert timing (e.g., crash avoidance versus crash mitigation).

The authors noted that the benefits of the FCW alert during surprise trials were restricted to tasks involving head-down glance activity and were not evident for the eyes-forward distraction tasks examined. Additionally, the researchers found that across the entire range of actual FCW alert and simulated FCW (via visual occlusion) conditions examined, a lack of both age and gender effects were generally missing. They concluded that this finding suggests that FCW alerts may be an effective means of equalizing a driver's abilities to avoid rear-end crashes.

Funke et al. (2007) explored the effects of stress, vehicle automation, and subjective state on driver performance and mood. A total of 168 participants completed a drive using a fixed-base driving simulator. Stress was manipulated by exposing drivers to a loss of control experience (i.e., a simulated winter drive that included periodic loss of control for the stressed driving condition, or no periodic loss of control for the non-stressed driving condition), while level of automation was varied across three levels (i.e., no automation/no lead vehicle, following a lead vehicle with a constant speed, and automated speed control that required the driver to perform lateral/steering maneuvers). Driver performance was measured through assessments of drivers' accuracy of vehicle control (i.e., variability in lateral position), attention to potential hazards (i.e., detection accuracy), and variations in mean speed that reflect an attempt to compensate for increased workload. Pre-experiment stress was measured through the Driver Stress Inventory (DSI; Matthews et al., 1997) while both pre- and post-task stress was measured through the Dundee Stress State Questionnaire (DSSQ; Matthews et al. 1999). Results indicated that both stress and automation influenced subjective distress, with higher levels of distress under the stressful driving conditions (i.e., the periodic loss of control in the simulated winter drive condition) and lower levels of distress under the automated speed control conditions; however, an interaction between stress and automation level was not found. Further, driver performance data revealed that vehicle automation impacted performance similarly in the stress and no-stress conditions. The findings suggested the potential benefits that the DSI screening tool may have for driver personnel selection efforts and the need to educate drivers, especially inexperienced drivers, about the negative consequences of stress.

Seto et al. (2008) conducted a field test to evaluate the effectiveness of a distance control assist system with an active accelerator pedal and a deceleration control. The system was designed to assist drivers in maintaining the following distance to a lead vehicle and relied on a laser radar sensor mounted near the front bumper and integrated with the main controller. The accelerator pedal was designed to react when the driver approached close to the vehicle ahead during car following (e.g., by generating force to push the pedal upward, applying brakes smoothly to decelerate the host vehicle) and when the system judged the driver's braking action was necessary. In the later condition, the system sounded a buzzer, flashed a caution message on the instrument panel, generated force for pushing the accelerator upward, and assisted with the braking action. Twelve drivers tested the system over an approximately 4-h drive totaling 113 km on ordinary roads, a restricted-access roadway, and expressways. Researchers examined order of system use (i.e., using on outbound but not reverse leg of the driving session and vice versa), time of day (morning or afternoon), and preferred following distance (short, medium, long). Data were gathered in regard to variables related to driving operations and vehicle behavior as well as variables related to the participants' mental and physiological conditions. The researchers found that the frequency and magnitude of deceleration by the driver decreased, and, from the characteristics of the time-to-collision distribution, the frequency of closing situations decreased, which may have contributed to the reduction of brake action by the driver. Through an analyses of drivers' Adapted Weighted Workload score on the NASA-TLX and the salivary amylase activity score, researchers found the system also resulted in decreased driver workload.

Vollrath, Schleicher, and Gelau (2011) explored the influence of cruise control and ACC on driving behavior. Using a motion-based simulator, 22 participants drove different scenarios in highway and motorway environments under three different conditions (i.e., assisted by cruise control, ACC, and manual driving without any system), with the inclusion of a secondary task condition. The findings suggest that, although drivers did not shift more attention to secondary tasks when driving with ACC, drivers did exhibit delayed reactions in critical situations (e.g., in a narrow curve or a fog bank). The authors concluded that additional research is needed to determine if the delay is due to reduced attention for signs and significant cues or to the effort involved in switching from automatic to manual control, with the ultimate recommendation of performing additional studies with experienced cruise control and ACC users in a naturalistic environment.

Using a motion-based simulator, Xiong, Boyle, Moeckli, Dow, and Brown (2012) examined ACC use patterns in a group of 24 participants deemed early adopters of ACC. They performed cluster analyses on drivers' use of ACC based on gap settings, speed settings, numbers of warnings issued, and ACC disengaged. Through these analyses and also drivers' subjective responses to questions regarding trust in ACC, understanding of system operations, and driving styles, they identified three groups of drivers based on driving behavior: risky behavior, moderately risky behavior, and conservative behavior. The results indicate that those drivers in the conservative group left greater following distances than the other two groups. Additionally, the risky drivers were found to respond later to critical events and to experience more ACC warnings. The authors conclude that safety implications are related to individuals' level of system understanding (e.g., greater awareness of the system resulting in what appears to be safer driving), level of trust in automation (e.g., greater trust resulting in greater cooperation with the system), and driving style and personality (e.g., greater overreliance on ACC with less understanding of the system and limitations).

Stanton, Dunoyer, and Leatherland (2011) evaluated in-car displays used to support Stop and Go ACC (S&G ACC). In an experiment with 12 automaker employees (who were unaware of the S&G ACC project), three different interfaces were proposed to support the detection of modal, spatial, and temporal system changes: an iconic display, a flashing iconic display, and a representation of the radar. The experiment was carried out using a host car equipped with S&G ACC, which the participant drove, and a leading vehicle driven by a member of the experimental team accompanied by a participant. Participants were asked to complete five driving tasks (i.e., follow at slow speeds, stop and start driving, lose lead on bend, lead brakes sharply, and lead cut-in) and to participate in a "multiple target identification test" to determine if drivers could correctly detect which of the objects the radar had identified as the leading vehicle. Results indicated that the speed of response was not influenced by the type of interface and that drivers correctly identified more changes detected by the system with the radar display than with the other displays; however, higher levels of workload accompanied this increased detection.

Bao, LeBlanc, Sayer, and Flannagan (2012) used data from the Integrated Vehicle-Based Safety System (IVBSS) naturalistic study to evaluate heavy truck drivers' following behavior in order to determine how a crash warning system influences their headway maintenance. Researchers followed drivers over a 10-month period. Two months of baseline period driving was followed by an 8-month treatment period, during which time drivers received warnings. Bao et al. found that the presence of warnings resulted in a 0.28-s increase of mean time headway with dense on-road traffic and a 0.20-s increase with wipers on. Additionally, drivers in the treatment condition responded to forward conflicts 15% faster than when in the baseline condition. Based on these findings, it was concluded that warning systems have the potential to increase heavy-truck longitudinal driving safety.

Saffarian, Happee, and de Winter (2012) studied drivers' headways in fog conditions in automated and manual driving scenarios. Using a driving simulator, 27 participants were tested using four scenarios: (1) clear visibility conditions and automated driving, (2) clear visibility conditions and manual driving, (3) fog and automated driving, and (4) fog and manual driving. In the clear and fog manual driving scenarios, participants used the primary vehicle controls (steering wheel, throttle, and brake pedals), while in the clear and fog automated driving conditions the driver only had to maintain lateral control over the vehicle (steering only, as the distance to the lead car was maintained through the automated controller). During each driving task, participants were asked to indicate their perception of current driving risk using a steering-wheel-mounted touchscreen. After each driving scenario, drivers were asked to complete the NASA-TLX (Hart & Staveland, 1988) workload assessment as well as four additional questions regarding perceptions of risk, driving safety, task ease, and confidence in ability to act. The results indicated that steering activity and perception of risk were elevated when the lead car was out of sight as opposed to when the car was in sight. They concluded that two advantages of maintaining close headway in fog are reduced perceived risk and improved lateral control. Additionally, no differences in risk perception were found between manual and automated driving when following distances were taken into account.

Lin, Hwang, and Green (2009) investigated the effects of time-gap settings and contents of secondary tasks on drivers' performance while reclaiming control from ACC in a car-following scenario of emergency braking by the lead vehicle. More specifically, Lin et al. explored time gaps preferred on expressway driving; which time-gaps maximized safety (as assessed by statistics of the gap when braking begins, brake pedal movement timefrom contact to 50% depression, the minimum gap during an encounter, and the number of crashes); and how time gaps were affected by the complexity of concurrent secondary tasks. Using a fixed-base bus-driving simulator, 30 professional bus drivers were observed as they drove using an intercity bus simulator within a highway traffic flow scenario with 12 random time-gap settings that ranged from 0.64 s to 2.4 s in intervals of 0.16 s. Three conditions were presented: no secondary task, simple secondary task (i.e., the addition of two numbers that were displayed on an LCD monitor at the right-hand side of the driver accompanied by a pre-question tonal cue), and complex task (i.e., a calculation task accompanied by a confirmation button press). Results indicated that the safer time gaps were longer than 1.60 s for non-secondary task distraction and longer than 2.08 s for being continuously distracted by secondary tasks.

Connected Vehicles

Nowakoski, Vizzini, Gupta, and Sengupta (2012) presented the final results of the USDOT SafeTrip-21 Initiative, which resulted in a real-time freeway end-of-queue alerting system that was tested along San Francisco Bay area freeways in California. The Networked Traveler Foresighted Driving Advanced Driver Assistance System was designed to promote drivers' situational awareness. Using vehicle-to-infrastructure (V2I) communications, drivers were provided with an auditory alert approximately 60 s before the driver reached slow traffic. Twenty-four drivers each drove instrumented vehicles for 2 weeks. During the first week of driving, baseline driving data was recorded; in the second week, the auditory alerts were activated and sounded when the driver approached traffic that was moving at least 24 km/h (15 mi/h) slower. The findings suggested that the system worked to increase driver situational awareness which resulted in smoother transitions into the end of the traffic queue, a reduction in the risk of rear-end crashes, and a small but significant reduction in mean peak deceleration rates during morning and off-peak travel as compared with the baseline conditions. Furthermore, drivers rated the majority of alerts good or neutral and reported that the system was most useful when traffic queues appeared unexpectedly.

SECTION 4. LESSONS LEARNED FROM OTHER DOMAINS

This section provides an overview of the findings from other fields that may be used to inform the development of automated driving systems. Findings have been drawn from the fields of aviation, rail, and process control.

Aviation and Unmanned Vehicles

An interesting parallel to automation in ground vehicles is the approach that different aircraft manufacturers have taken in flight automation. Modern aircraft offer many automated components, from auto-throttles and anti-skid braking systems, to flight management systems

(FMS) that have largely negated the need for most flights to carry a dedicated flight engineer and navigator. However, control automation – that automation that assists the pilot or assumes some level of control over the aircraft – presents some interesting challenges in terms of human-automation interaction. These challenges are perhaps best illustrated by the different philosophies of two major aircraft manufacturers: Boeing and Airbus.

Orlady and Barnes (1997) summarized these differences. Boeing's philosophy on aviation automation is that the pilot is the final authority in flight operations and that automation should assist rather than replace the pilot. In contrast, the Airbus philosophy is that automation should enhance aircraft and system performance and maintain the aircraft's flight envelope by not working against the operator unless necessary for safety. This difference in how the human-automation interaction is managed, is subtle, yet critical. While the Boeing automation philosophy allows for the pilot to commit errors, lapses, and mistakes that can lead to incidents, the Airbus philosophy can prevent a highly trained pilot from executing a maneuver needed for safety.

Two crashes illustrate the respective problems with each approach. In December 1995, a Boeing 757 en route to Cali, Columbia, crashed into a mountain (Aviation Safety Network, 2013a). The resultant investigation determined that one of the contributing factors was the pilot's failure to retract the speed brake during an attempted climb. In a more automated-type system (e.g., Airbus) the automation would have automatically retracted the speed brakes. In the Boeing system, the pilot's lapse helped precipitate the crash. The second incident took place in June 1988, when an Airbus A320 was being demonstrated at an airshow at Mulhouse-Habsheim Airport in France. As part of the airshow, a low- speed flyover was to occur. Instead of the planned 33-m flyover, and due to a number of factors, the pilots executed the maneuver at approximately 9-m. However, as the plane passed below approximately 15 m, the automated landing system took control of the plane and crash landed into the trees surrounding the field. In this case, the automation functioned as intended, but outside the intentions of the pilots.

Although both manufacturers continue to pursue greater levels of sophistication in cockpit automation, increased automation can introduce the possibility of errors that previously did not exist. Mode confusion and not fully understanding the actions and capabilities of the automation can lead to errors. However, the proper use of automation can also result in degraded performance. Sumwalt, Thomas, and Dismukes (2002) noted that a pilot's monitoring behavior is degraded during periods when the pilot is not actively flying, a situation enabled by automation technology. Likewise, research has indicated that pilots' manual flight skills are degraded by longer and more frequent use of automation (Wiener, 1988; Sherman, 1997). As a result, a recent Safety Alert for Operators (SAFO) issued by the Federal Aviation Administration (FAA, 2013) encouraged pilots to use manual (non-automated) control of the plane when possible, as the continuous use of autoflight systems leads to a lack of reinforcement of pilot skills.

This fact is of great interest in the domain of automated ground vehicles. Defining intent, and ceding ultimate control, is associated with some risks in L2 and L3 automated vehicles. It is conceivable that automated components within the vehicle will be designed to account for a certain amount of human variability. That is, people will make mistakes, and many of these will likely be understood. The designs of these automated components will account for such mistakes and act accordingly. However, the risk in this approach is that the automated component does not correctly account for the intentions of the driver and acts inconsistently with his or her expectations. Thus, the designers of highly automated components can provide ultimate decision authority to the vehicle automation, or they can provide ultimate authority to the driver. In the case of ultimate control and authority to the vehicle automation, the electronics can prevent a driver from making a mistake (e.g., falling asleep and failing to steer the vehicle). In the latter case, the driver can detect a rare event and evade accordingly where the automated component may not have been able to adequately adjust. Again, further research into shared authority and transitions between the driver and automation could provide significant benefit towards understanding and accounting for these risks.

Additionally, questions of pilot overreliance (and the associated skill degradations from such overreliance on automation) raise questions of similar effects in drivers. Could there be a risk of overreliance on the automation when drivers are able to use an L2 or L3 automated driving system for some extended period of time and, if so, for how long? This risk comes from the potential for drivers to constantly seek to activate the automated driving component of a vehicle after becoming familiar with the automation and how it functions. Over time, this can possibly lead to a degradation in driver skill, as the reinforcement coming from constant engagement in the driving task is now lacking. This issue is of concern in the aviation domain, where pilot skill has been observed to suffer in the presence of frequent and continuous use of automated functions.

This section provides an overview of key recent human factors findings to be drawn from the field of aviation. Several have provided overviews of the issues associated with aviation automation (e.g., Finn & Scheding, 2010; Kaber & Prinzel, 2006). Additional efforts have looked at:

- System-wide trust assessments;
- Characterization of adaptive systems, generally, and the benefits of co-adaptive aiding;
- Effects of observability, auditory cueing, and conflict probe automation;
- Issues associated with the use of unmanned ground vehicles (UGVs) and unmanned air vehicles;
- Examination of workload and stress metrics; and
- Workload associated with the implementation of conflict detection and resolution (CD&R) systems.

Overviews of the issues associated with automation within the field of aviation are numerous (Sheridan, 2006). For example, Finn and Scheding (2010) provided an overview of the developments and challenges for automated unmanned vehicles. Langan-Fox, Sankey, and Canty (2009) conducted a critical review of the research pertaining to the measurement of human factors issues in current and future air traffic control. They predicted that, given the changing role of air traffic controllers and of airspace requirements and configurations, issues of stress, trust, and boredom will become more significant. Landry (2009) provided a review of flight deck automation with an emphasis on examples and design principles. Landry discussed human factors, integration, safety, and certification issues associated with the development of automated flight deck systems. Specific attention was given to the challenges facing those who will design future systems.

Additionally, Kaber and Prinzel (2006) presented a critical review of the literature associated with approaches to adaptive and adaptable task/function allocation and the adaptive interface technologies for effective human management of complex systems that were likely to be issues for the Next Generation Air Transportation System and the focus of research under the Aviation Safety Program, Integrated Intelligent Flight Deck Project. Kaber and Prinzel noted that their main finding was that adaptable system designs requiring human delegation of task and function authority to automation during performance may increase operator workload, which would then lead to degradations in situational awareness and performance. In regard to adaptive interfaces, they noted the need for consideration of multiple user characteristics, preferences, and behaviors in the design process. Further, they observed that adaptive interface features need to be linked to specific task requirements and that it is necessary to maintain some consistency in interface design across modes of system operation so that operators can effectively apply mental models.

Feigh, Dorneich, and Hayes (2012) presented a systemic framework characterizing adaptive systems (i.e., those systems that can appropriately modify their behavior to fit the current context). They presented a two-part framework that categorizes ways in which adaptive systems can modify their behavior as well as trigger mechanisms through which adaptive systems can sense the current situation and decide how to adapt. The taxonomy of system modification included the following:

- Modification of Function Allocation: dynamically changing who (human or machine) performs each task (includes task sharing and task offloading).
- Modification of Task Scheduling: dynamically changing when tasks are performed, including task duration and priority.
- Modification of Interaction: dynamically changing how the system interacts with the user (includes system style, interface features, modality, and amount).
- Modification of Content: dynamically changing what information is presented to users, including what categories are presented and at what level of detail or abstraction.

The taxonomy of triggers presented is based on information that can be sensed, observed, or modeled to create an understanding of context relevant to the adaptive system's decision-making. The proposed taxonomy is divided into five categories:

- Operator-based triggers: adaptations are initiated by the operator or by a system-assessment of the operator state.
- System-based triggers: adaptations are triggered by predicted system states or different modes of system operations.
- Environment-based triggers: adaptations are triggered by the environment or external events.
- Task- and mission-based triggers: adaptations are based on a coherent set of goals or sub-goals and/or task accomplishments.
- Spatiotemporal triggers: adaptations are triggered by either time or location.

The authors envision this framework as serving as a starting point for system designers.

Christensen and Estepp (2013) expanded upon the theory of adaptive aiding by measuring the effectiveness of co-adaptive aiding, wherein they allowed for both the system and the user to adapt to each other. They compared the physiological activation of adaptation with manual activation or no activation of the same automation and cueing systems using 10 participants and a PC-based supervisory mission control scenario. Three experimental trials (one each of no aiding, physiologically activated aiding, and manual activated aiding) were conducted. The tasks were meant to represent future operator control tasks and to tap into a wide range of cognitive skills including working memory, visual search, object recognition, task switching, and flexible management of conflicting priorities. Physiological data, including that for EEG, electrooculography (EOG), and electrocardiogram (EKG) were recorded. Results demonstrated that in the first 2 days of testing no significant differences in performance were noticed between conditions. However, on the third day, physiological adaptation produced the highest performance. The researchers concluded that the extended testing period provided enough time and experience for user adaptation as well as online system adaptation, which demonstrated co-adaptive aiding. They concluded that the results may be used to implement more effective adaptive workstations in a variety of work domains.

Niederée et al. (2012) noted that little research has been completed on the effect of observability (how well internal states may be inferred from external outputs), mood states, and low workload conditions on human errors in handling automation in the field of highly automated aircraft. This research attempted to address that gap. Assumptions were made regarding observability, mood states (i.e., tension, activation, positive mood or happiness, achievement potential, and extraversion), and workload. First, lacking observability of the automation would result in an increase in human errors in handling automation. Second, mood state assumptions included:

- Tension: higher tension levels were posited to be related to lower levels of trust, which in turn would impact human-automation cooperation negatively, thus resulting in a greater number of human handling errors.
- Activation (i.e., the level of drive and energy): a higher level of activation would result in a higher number of handling errors.
- Positive mood or happiness: fewer errors would occur when the mood was more positive.
- Achievement potential (i.e., the operator's perceived capability of achieving a high level of performance due to a high level of alertness and concentration): high potential would result in a lower level of handling errors as a result of better concentration and alertness.
- Extraversion (i.e., a disposition to behave impulsively): a higher level of extraversion would result in a larger number of handling errors.

Third, higher workload was expected to reduce handling errors. To test these assumptions, 24 participants were required to monitor aircraft automation to keep the aircraft's pitch angle within certain boundaries and to keep the engines functioning as expected. When the automated pitch was outside boundaries, participants had to correct the automation by pressing a button. Engine function monitoring was a secondary task and required participants to make adjustments when repeated values (associated with engine speed, exit fuel temperature, or fuel flow) varied. Workload associated with the engine monitoring was varied between the monitoring of three variables versus the monitoring of one variable (exit fuel temperature). After each task, participants rated their mood. The results indicated that significant effects of automation's observability and the participants' level of positive mood and extraversion on the number of handling errors were present. Researchers considered handling errors to be indicators for communication breakdowns as they were the result of the automation insufficiently informing the operator on its actions. They concluded that the results highlighted the need to focus on automation's observability when designing highly automated systems, especially in safety-critical domains, and on the need for further analysis of the effects of mood states on human-automation interaction.

Dehais, Causse, and Tremblay (2011) assessed the efficacy of cognitive countermeasures based on the technique of information removal to enhance human operator attentional disengagement abilities when facing attentional tunneling. They noted that conflicts with automation lead to the degradation of operators' performance by promoting excessive focusing on a single task to the detriment of the supervision of other critical parameters. To test their cognitive countermeasures, they developed an experimental design that incorporated an actual UGV, a ground station, and a computer interface dedicated to triggering special hazards within the scenario. Participants were asked to complete a target localization and identification task. Authority conflict was introduced by a low-battery event at a point when participants were deeply involved in the target identification task. The system was designed so that this hazard resulted in the robot returning to base autonomously. Participants were warned of this development through three visual alerts which displayed on the user interface. The cognitive countermeasure altered the screen display and superimposed conflict explanation information over the display. Researchers tracked participants' decision-making at the time of the failure, ocular activity (to ensure that attentional shrinking was mitigated as indicated by increased saccadic activity and a greater number of scanned areas of interest), and heart rate (to establish whether sympathetic activity was reduced because of the countermeasure). Findings suggested the use of cognitive countermeasures appeared to be effective at mitigating excessive focus issues in the UGV environment. They concluded that the principle of cognitive countermeasures could be applied to a large domain of applications that involve human operators interacting with critical systems.

Kaber and Kim (2011) investigated the effects of advanced auditory cueing of control mode changes in an adaptively automated system on human performance and explained cognitive behaviors at mode changes by using a computational cognitive model. More specifically, they assessed whether cueing of mode transitions would reduce the return-to-manual control (performance) deficit effects associated with the mental model changeover (i.e., switching from long-term memory into working memory). They hypothesized that increased long-term memory and working memory transactions would degrade operator performance, specifically response time and accuracy. To explore these effects, a dual-task piloting simulation was developed that involved tracking and tactical decision- making. Performance data was collected based on auditory cueing or no cueing of the mode transitions in the tactical task. Kaber and Kim noted that the human performance data did not reveal any differences between cued and non-cued trials, possibly because of distraction from the tracking (secondary loading) task. These findings were compared to a computational GOMS (goal, operators, methods, and selection) model. This model was then refined to assume that memory stores are used on an ad hoc basis after high-workload mode transitions and with consideration of human parallel processing in dual-task performance; the refined model was determined to have greater plausibility for representing user behavior. The authors concluded that the use of memory stores for controlling an adaptive system provides insight into the impact of the cueing of mode transitions which can inform future system designs.

Rovira and Parasuraman (2010) examined how the type of automation imperfection (miss vs. false alarm) would affect air traffic service providers' (ATSPs') performance and attention allocation. The experiment was conducted using a head and eye tracker, as well as an air traffic control (ATC) simulator. An automated conflict probe tool highlighted aircraft projected to be in conflict within the next 6 min. The conflict probe warning remained visible until the initial point of loss of separation or until the ATSPs performed resolution responsibilities. Twelve performance-level ATSPs served as participants, who were presented with four levels of automation support (manual performance, reliable automation, miss automation, and false alarm automation) across four 30-min scenarios. Conflict resolution, resolution performance, eye movements, and ratings of trust and self-confidence were measured. Results indicated that ATSPs detected conflicts faster and more accurately with reliable automation than with manual performance. They concluded that when the primary task of conflict automation was automated, even highly reliable yet imperfect automation (miss or false alarm) resulted in serious negative effects on operator performance. However, Rovira and Parasuraman cautioned that the further in advance a conflict probe automation predicts a conflict, the greater the uncertainty of prediction; therefore, they urged designers to provide users with feedback on the state of the automation or other tools that allowed for inspection and analysis of the data underlying the conflict probe algorithm.

Chen, Durlach, Sloan and Bowens (2008) examined the ways in which human operators interacted with simulated semi-automated UGVs, semi-automated unmanned aerial vehicles (UAVs), and teleoperated UGVs (Teleop). Additionally, operators' span of control of robotic assets was studied. During the experiment, participants were given either a single robotic asset (i.e., UAVs, UGVs, or Teleop) or all three. Chen et al. (2008) found that target detection performance was lower with three than with a single unmanned vehicle (UV), and that participants were also less likely to complete their missions in the allotted time. Chen and Terrence (2009) investigated the performance and workload of the combined position of gunner and robotics operator in a simulated environment to determine how aided target recognition capabilities for the gunnery task with imperfect reliability (false-alarm-prone versus miss-prone) might affect the concurrent robotic and communications tasks.

Chen and Barnes (2012) studied the effects of an intelligent agent on the performance of robotics operators. Specifically, they looked at the effects of imperfect automation and individual differences on the management of a team of ground robots. The intelligent agent was manipulated so as to perform perfectly or to be either false-alarm or miss prone. Results indicated that when the intelligent agent was reliable, it was helpful in reducing overall mission times. Additionally, they found that the type of imperfection (false-alarm vs. miss prone) affected operators' performance of tasks involving visual scanning (i.e., target detection, route editing, situational awareness). Further, they noted a consistent effect of visual density (i.e., clutter of the visual screen) for multiple performance measures. They also discovered that participants' attentional control and video gaming experience affected their overall multitasking performance and that participants with greater spatial ability consistently outperformed low-spatial-ability counterparts in tasks that required effective visual scanning. The authors concluded that while intelligent agents can benefit the overall human-robot teaming performance, the effects of type of agent unreliability, tasking requirements, and individual differences needed to be considered in light of their complex effects on human-agent interaction.

Chen and Terrence (2009) also examined whether performance was influenced by individual differences in spatial ability and attention control. They found that within the context of a simulated military multitasking environment, false-alarm-prone alerts were more detrimental to those with higher perceived attentional control while miss-prone automation errors were more harmful for those with lower perceived attentional control. Additionally, participants with low spatial ability preferred visual cueing while high-special-ability participants favored tactile cueing. These findings can be used to inform personnel selection for robotics operations, robotics user interface designs, and training development.

Neyedli, Hollands, and Jamieson (2011) evaluated display formats for an automated combat identification aid. They tested four visual displays that showed both target identity and system reliability information while manipulating display type (i.e., pie, random mesh) and display proximity (i.e., integrated, separated) of identity and reliability information. Two experiments were conducted. In the first, participants used the displays while engaging targets in a simulated combat environment. In the second, participants briefly viewed still scenes from the simulations. The researchers concluded that the integrated display format and a random mesh display were the most effective displays tested. They recommended the use of this format and display to indicate identity and reliability within an automated combat identification aid system.

Cummings, Mastracchio, Thornburg, and Mkrtchyan (2013) explored boredom within operators of UAVs. Cummings et al. conducted a long-duration, low-task-load experiment (i.e., three task loads requiring input over 10, 20, or 30 min) using a multiple UV simulation environment. They found that participants' reaction times to system-generated events and their ability to maintain directed attention decreased over the 4-h test period. Additionally, participants spent almost half the study time in a distracted state. The results led to the conclusion that distraction due to boring, low-task-load environments could be managed through efficient attention switching; however, future work is needed to determine the frequency and duration of attention state switches.

Parasuraman, Cosenzo, and Visser (2009) explored the use of automation to support human operators as they supervised multiple UAVs and UGVs. Two experiments were conducted to determine the efficacy of adaptive automation in a simulated high-workload reconnaissance mission that involved four subtasks: target identification; UGV route planning; communications, with embedded verbal situation awareness probes; and change detection. The researchers used a custom-developed simulation, the robotic non-commissioned officer, to isolate some of the cognitive requirements associated with a single operator controlling robotic assets within a larger military environment. The first experiment was a baseline study without automation that allowed researchers to determine the levels of low- and high-task load used in the second experiment. Task load was varied through the manipulation of the difficulty of the UAV and communications task at each of two levels in a 2×2 factorial design. Tasks were presented in separate windows of a computer monitor along with a separate window displaying a situational map of the reconnaissance area. The second experiment compared three automation conditions: manual; static automation, in which an automated target recognition (ATR) system was provided for the UAV task; and adaptive automation, in which individual operator change detection performance was assessed in real time and used to invoke the ATR only when change detection was below a threshold. Although participants in both automated scenarios had greater change detection accuracy and situational awareness and lower workload than in the manual scenario, the greatest beneficial effects resulted from the adaptive automation scenario. The results demonstrated the efficacy of adaptive automation for supporting human operators tasked with supervising multiple UAVs and UGVs under high-workload conditions.

Guznov, Matthews, Funke, and Duke (2011) evaluated whether the RoboFlag simulated environment was capable of eliciting dissociations between performance, workload, and subjective stress responses through the manipulation of environmental uncertainty (i.e., uncertainty about key elements of the operating environment) and maneuverability (i.e., controllability of the vehicle's trajectory). An experiment with 64 participants indicated that the RoboFlag task resulted in increased workload and elevated distress. Additionally, they confirmed the benefits of a multivariate approach to assessment. The authors also found that, contrary to their expectations, distress and some aspects of workload were highest in the low-uncertainty condition, which they concluded may suggest that overload of information may be an issue for UAV interface designers.

Ruigrok and Hoekstra (2007) reported on human factors evaluations of free flight (i.e., operators, under instrument flight rules, have the freedom to select their path and speed in real time). As background, Ruigrok and Hoekstra provided detailed results from previous human-in-the-loop experiments conducted in a flight simulator with specific attention given to the human factors issues, particularly workload. The Airborne Separation Assurance System (ASAS) in the free flight experiments consisted of:

- Automatic Dependent Surveillance – Broadcast (ADS-B) system which provides a digital link between aircrafts so that they can "see" each other. This system also includes the Traffic Information Services – Broadcast (TIS-B) system that contains similar data to the ADS-B but is generated by ground stations using radar.
- Cockpit Display of Traffic Information (CDTI) on Navigation Display (ND). Pilots are provided with call sign, relative/absolute altitude, groundspeed/calibrated airspeed, track, and a vertical speed arrow.
- Conflict Resolution and Detection (CR&D) algorithms. The CR&D function is based on aircraft *state* information and incorporates position plus three-dimensional speed vector information from the aircraft's own position and other traffic. Conflicts are detected up to 6 min ahead of the aircraft. If the closest point of approach is below the separation standard (5 mm horizontally and 950 ft vertically), a conflict is shown.
- Predictive ASAS (PASAS). Initial experiments suggested that an ASAS system should be capable of conflict prevention, i.e., not introduce conflicts while maneuvering the aircraft. PASAS calculates which headings and vertical speeds will result in a conflict with another aircraft within the look-ahead time (6 min). Potential conflicts are displayed on the primary flight display and ND.
- Alerting Logic. Distinctions are made between potential conflicts within 0–3 min away from intrusion and 3–6 min away from intrusion. Pilots are given rules for acceptable actions within each conflict zone.

The 1997 human-in-the-loop experiment varied the traffic density, level of automation, and nominal (i.e., normal)/non-nominal conditions (i.e., special events including ASAS failures and pilots not obeying the rules). Three levels of automation were used for resolution activation via the aircraft autopilot:

- Manual: no automation: use normal select mode for heading, vertical speed, and/or horizontal speed.
- Separate: two modes available: vertical or horizontal maneuver.
- Combined: one mode performs combined horizontal and vertical maneuvers.

Researchers tested combinations of traffic densities and levels of automation in both nominal and non-nominal conditions. A second human-in-the loop study explored human factors issues associated with future air traffic management (ATM) systems, covering transitions in time (mixed equipage, i.e., aircraft equipped for free-flight operations and aircraft not equipped for these operations) and transitions in space. Follow-on experiments explored what would happen when more aircraft were controlled by humans (versus the more ideal automatically controlled traffic) and how low flights could go when flying into an airport in free flight until the final approach fix.

Results indicated that pilots' subjective workload increased slightly but remained within acceptable levels, even in high-traffic densities. To provide pilots with additional situational awareness, further studies were conducted which incorporated an intent-based ASAS (in contrast to the aforementioned state-based ASAS). While the conflict can be calculated based on aircraft state (i.e., ground speed, track, and vertical speed), conflicts can alternatively be calculated based upon aircraft intent (i.e., flight plans). The intent-based CD&R researchers implemented priority rules instead of cooperative solutions, meaning that one aircraft was assigned priority while the other was expected to maneuver. Studies developed a human operator model for fast-time simulations based on measurements during the simulation and obtained pilot feedback on the use of intent-based ASAS information in the en- route flight phase. A follow-on study validated outcomes from the fast-time simulations and extended the scope of the flight-to-arrival and -departure phases. An additional study, the Mediterranean Free Flight study, combined state-based ASAS with the priority rules associated with intent-based ASAS. Based on these studies, researchers determined that although the intent-based CD&R system resulted in reduced pilot workload, it also resulted in increased complexity, the need for priority rules, and the resolution of potential compatibility problems between different flight management systems and large bandwidth requirements. Additionally, the intent-based system was not effective at solving multi-aircraft conflicts. The state-based system was simpler, required less bandwidth, was easier to retrofit, and could solve multi-aircraft conflicts in parallel. The costs were higher pilot workload in similar circumstances and a smaller look-ahead time, which resulted in less efficient maneuvers. The optimal CD&R was suggested to be the state-based CD&R with the addition of intended or target flight level.

More recently, Bonini, Dupré, and Granger (2009) presented findings associated with the En Route ATM Soft Management Ultimate System (ERASMUS) project, a project proposing to share air traffic controllers' workload with automation in order to increase capacity. Bonini et al. tested the potential of ERASMUS in a demonstration with operational experts. The underlying assumption of the ERASMUS project is that decreasing the attention load of the controller through automation reduces the number of conflicts, giving the controller more time and resources to control a greater number of aircraft. ERASMUS provides conflict detection (based on aircraft trajectory prediction) and conflict resolution (e.g., minor speed changes, new flight levels or trajectories). The system can function autonomously or with controller input. A demonstration was conducted over a period of 3 days and included the ERASMUS autonomously working as a support system for the meta-sector planner and as a support system for the tactical controller. The results indicated that participants found that the HMI was intuitive and attractive, illustrated the concepts effectively, and allowed them to envision their use in an operational setting. Additionally, users were more receptive to the system for its detection potential versus its conflict resolution abilities.

Rail

Recent rail-related automation studies examined the importance of managing human factors in safety- critical fields, how automation contributed to sustainability efforts, the role of automation in rail- signaling efforts, and the effects of automation on workload. Each of these concepts is discussed further within this section.

Cunningham (2007) discussed the importance of managing human performance within safety-critical fields, noting that rail operators are implementing lessons learned from other fields, such as aviation. He noted that there are two ways to manage human performance: through technology (i.e., the automation of control and operations) or through human resource management (i.e., training individuals and implementing processes that address the psychological aspects that cause human error). Within the rail industry, many efforts have been implemented to manage human performance, including psychometric testing to assess individuals' ability to perform repetitive and monotonous tasks, ensuring limited workdays and regular breaks, and engaging staff in fatigue-education programs. In regard to automation, the Automatic Warning System and the newer Train Protection and Warning System (which assumes control and applies the brakes in cases where the driver is distracted or makes an error) have been proven to minimize human error.

Vlad and Tatarnikov (2011), in their examination of sustainable rail transportation efforts, discussed Vancouver's SkyTrain Driverless Urban Rail System. They noted that because the system is automatically controlled, system safety and efficiency is increased because the system can operate longer and under more optimal conditions than if it were manually driven. However, they cautioned that, within this system, operators are still affected by fatigue which results in suboptimal parameters of the rides (i.e., acceleration, braking, stopping) and, in extreme cases, could lead to accidents with potential health, safety, and environmental consequences.

Balfe, Wilson, Sharples, and Clarke (2011) conducted a qualitative study involving 10 operational rail staff who regularly used automated systems in rail signaling. Specifically, Balfe et al. (2011) looked at the most advanced form of automation in use in the UK rail network, automatic route setting (ARS), which is used in 12 signaling centers across the UK where it has been implemented on Visual Display Unit (VDU)-based signaling systems. The ARS system sets appropriate routes and attempts to deal with conflicts. The system attempts to account for both static (e.g., train service pattern) and dynamic (e.g., speeds of the trains) properties, but signalers retain the ability to take manual control of the system. Ten semi-structured interviews were conducted to develop an understanding of the role of automation in UK rail signaling and how the design influences operators' use of and interaction with the automation. The findings indicate that, despite having generally low levels of understanding about the system and an inability to predict automation system actions, signalers have developed coping mechanisms that enable them to use the technology effectively. Additionally, drawing from the literature and the interviews, 10 key principles of automation were identified that are intended to address concerns associated with introducing and designing automated systems (i.e., reliable, competent, visible, observable, understandable, directable, robust, accountable, proactive control, and skill degradation; p. 49).

Working with Network Rail (the rail infrastructure owners in the UK), Pickup, Wilson, and Lowe (2010) described the development of the Operational Demand Evaluation Checklist (ODEC): a tool for determining mental workload in signal operators that has potential applications in other domains. Development of the ODEC used a Repertory Grid Technique (RGT) to identify the most relevant elements of the signaling system and each element's potential to result in high- or low-workload situations. Data from the RGT phase were then transferred to a tool that could record data on each element into the system, the output of which would enable one to determine the extent to which the signaling system contributed to signaler workload. The system was further refined to include additional data considerations. Additionally, researchers validated the system, in part through a comparison of ODEC output data with the rail industry grading system which was used for human resource management purposes. They noted that although the tool was not developed for predictive use, ergonomists have used the tool to assess the current and potential demands on a system and workload as the result of technology changes and that the RGT method could be replicated in other domains or organizations.

Control System Automation

Human control of a number of different systems may be aided by automation. Automation can assist by providing physical control or by providing decision-making aids to users. The work discussed in this section deals with topics such as reliance, countermeasures, and the different approaches to presenting the status of automation; all of these topics have a great deal of importance for the overall success of human-automation interaction.

McBride et al. (2010) studied whether older and younger adults differed in the way they used automation a well as whether their subjective experiences of interacting with the automation differed, as measured by trust and workload ratings. A group of older ($M = 70.17$, $SD = 2.8$) and younger ($M = 20.39$, $SD = 2.39$) participants used a simulated dual-task scenario (i.e., accepting inventory and dispatching trucks). One of the tasks (i.e., dispatching trucks) was supported by an automated aid that was reliable 70% of the time. They found that younger adults outperformed older adults in both tasks. Additionally, when the automation was incorrect, younger adults exhibited less dependence on the automation and took significantly less time to verify the automation's suggestion than did older adults. Findings also suggest that older adults reported greater trust in the automation and higher workload compared to younger adults.

Jipp (2012) studied the extent to which individual differences in fine motor abilities affected indoor safety and efficiency on human wheelchair systems. The purpose of this effort was to reduce the large number of indoor wheelchair accidents through the use of assistance systems that incorporated a high level of automation. Through testing of participants' aiming, precision, and arm-hand speed abilities as they navigated a specially designed course, he found that participants with lower fine motor abilities had more collisions and required more time for reaching goals. He concluded that adapting the wheelchair's level of automation to these fine motor abilities could improve indoor safety and efficiency. Moreover, these findings suggest the need for further examinations of the impact of individual difference on the design of automation features for not only powered wheelchairs but also for other applications of automation.

Röttger, Bali, and Manzey (2009) explored the impact of automation on operator workload and behavior in process control fault management using subjective, cardiovascular, and secondary task performance indicators. Twelve participants were asked to detect a malfunction in a simulated life- support system of a space capsule, identify the source of the malfunction, and maintain appropriate atmospheric conditions through manual interventions for the duration of the malfunction. Three levels of automation were tested: some structured guidance for fault diagnosis, suggested diagnosis and recommendations for fault management, and automated management of system if not vetoed by user. Results indicated that better performance (i.e., shorter fault identification times) occurred in the medium to high automation scenarios, that control activity increased considerably during system malfunctions especially in the manual and mid-range automation levels; that during fault-free phases no differences in information sampling actions and control actions occurred between automation conditions, but significantly more control actions were observed in the manual condition; and that subjective workload decreased almost linearly with the increasing degree of automation. The results suggested that the differences between the performance and workload effects of the different levels of automation were slight. As a result, they were unable to draw definite conclusions about the relative advantages of each system in terms of real-world implementation and noted that additional factors, such as system reliability and monitoring complexity, may need to be considered.

Merritt, Heimbaugh, LaChapell, and Lee (2012) examined the influence of implicit attitudes toward automation on users' trust in automation. They noted that although previous empirical work examined explicit (i.e., conscious) influences on user level of trust in automation, implicit influences had not yet been measured. To address this gap, the authors examined concurrent effects of explicit propensity to trust machines and implicit attitude towards automation on trust in an automated system. The differential impacts of each were examined under varying automation performance conditions (i.e., clearly good, ambiguous, clearly poor). Participants were asked to complete a self-reported propensity to trust and an Implicit Association Test measuring implicit attitudes towards automation. Participants were then asked to complete a simulated X-ray screening task wherein automation performance was manipulated within subjects by varying the number and obviousness of errors. They found that propensity to trust and implicit attitudes toward automation did not significantly correlate. Further, results indicated that when the automation's performance was ambiguous, implicit attitude significantly affected automation trust, and its relationship with propensity to trust was additive. That is, increments in either were related to increases in trust. Additionally, they found that when errors were obvious, a significant interaction between the implicit and explicit measures was found: those high in both implicit and explicit measures had higher trust. They concluded that implicit attitudes have important implications for trust; that users may not be able to accurately report why they experience a given level of trust; and that measurements of explicit and implicit predictors may be necessary to understand why users trust, or fail to trust, automation.

Merritt (2011) examined the influence of user moods and emotions on the reliance on automated systems. Following the presentation of video clips selected to induce a positive or negative mood, participants were instructed to interact with a fictitious automated system on an X-ray screening task while trust, liking, perceived machine accuracy, user self-perceived accuracy, and reliance were assessed. These variables, along with propensity to trust machines and state effect, were included in a structural equation model. Results indicated that happiness significantly increased trust and liking for the system throughout the task. Additionally, liking was the only variable that significantly predicted reliance early in the task, whereas trust predicted reliance later in the task. Perceived machine accuracy and user self-perceived accuracy had no significant direct effects on reliance. Based on these findings, Merritt noted that liking a new system may be key to appropriate automation reliance. Additionally, positive affect can be easily induced and may be a lever for increasing liking.

Lyons and Stokes (2012) considered human-human reliance during a computer-based scenario where participants interacted with a human aide and an automated tool simultaneously (noting that although previous research explored the differences in human-human versus human-machine reliance, few studies have examined such reliance when individuals are presented with divergent information from different sources). To assess these differences, the authors conducted a convoy leader simulation. Forty participants were asked to decide which of three routes could be used to safely send convoy operations. To aid in their decision-making, participants were provided with three sources of information: route parameters, an automated tool that highlighted historical enemy hostile threats, and a human aide who suggested a route based on intelligence reports of hostile activity. Risk was manipulated through the human aide, who provided probability of attack as being low, moderate, or high, and by information from the automated tool that did not support the guidance offered by the human aide. Participants' human reliance intentions, automation reliance intentions, human behavioral reliance (i.e., whether or not the participant took the route suggested by the human aide), and manipulation checks (i.e., participants' assessments of perceived risk) were assessed. The findings suggest that when individuals are provided information from both a human aide and automation, their reliance on the human aide decreased during high-risk decisions. This information is useful in that it adds to an understanding of biases and preferences that exist during complex human-human and human-machine interactions.

Reichenback, Onnasch, and Manzy (2011) investigated how human performance consequences of automated decision aids are affected by the degree of automation and the operator's functional state. They asked participants to perform a simulated supervisory process control task with one of two decision aids providing support for fault identification and management. The process was simulated using a "microworld" simulation that simulated an autonomously running life-support system consisting of five subsystems critical for the maintenance of atmospheric conditions (AutoCAMS 2.0; citing Manzy et al., 2008). Participants were required to supervise the five subsystems, including the diagnosis and management of system faults. Depending on the fault involved, participants either performed the fault diagnosis and management manually or with the assistance of one or two different versions of an automated aid. In the first, an alarm was accompanied by a message providing the specific diagnosis for the fault; in the second, the system also implemented all necessary steps autonomously if confirmed by the participant. Tests were conducted over the course of two sessions. One took place during the day, and the other took place during the night after a prolonged waking phase of more than 20 h (extreme evening types were excluded based upon the results of a Morningness-Eveningness Questionnaire, citing Griefahn, Künemund, Bröde, & Mehnert, 2001). The results showed that decision aids could support humans effectively in maintaining high levels of performance, even in states of sleep loss, with more highly automated aids being more effective than less automated ones. Additionally, participants suffering from sleep loss were found to be more careful in their interaction with the aids (i.e., less prone to effects of complacency and automation bias). However, the authors cautioned that costs arose that included a decline in the secondary-task performance and an increased risk-of-return-to-manual performance decrements. They concluded that automation could help protect performance after a period of extended wakefulness. Also, operators suffering from sleep loss seemed to compensate for their impaired functional state by reallocating resources and showing a more attentive behavior towards possible automation failures (Reichenback et al., 2011).

Key Lessons Learned from Other Domains

The outcomes of these studies from the fields of aviation and unmanned vehicles, rail automation, and control system automation can be used to inform the design and development of automated vehicle systems moving forward. As noted, there are many lessons to be learned from the aviation and unmanned vehicle industry regarding the sharing of authority and the problems associated with overreliance on automated systems (and the associated skill degradations from such overreliance on automation). In regards to design, researchers have recommended that adaptive interface features be linked to specific task requirements and that some degree of consistency in interface design across modes of system operation be maintained so that operators can effectively apply mental models (Kaber and Prinzel, 2006). When dealing with imperfect systems, even in highly reliable systems, imperfect automation (e.g., missed alarms, false alarms) can result in serious negative effects on operator performance (Rovira & Parasuraman, 2010). Moreover, the effects of imperfect automation can vary based on differences among individuals (Chen & Barnes, 2012; Chen & Torrence, 2009). Finally, distraction due to boring, low-task-load environments may be effectively managed through efficient attention switching; however, more research is needed to determine both the optimal frequency and duration of state switches (Cummings et al., 2013).

The findings from the rail automation studies may be used to inform the integration of automated vehicles, both group rapid transit and personal rapid transit, into new or existing public transit networks. Automated systems could be used to improve safety and efficiency of transit systems. System service hours could also be extended as automated transit alternatives can operate longer than those that are manually driven. While potentially beneficial, the integration of automated transit alternatives will require parallel efforts by human resource managers, which are designed to educate operators and to minimize fatigue and boredom resulting from monotonous and repetitive tasks.

These control system automation studies suggest the importance of considering variations among users in the design of the system. For example, younger and older users will rely on automation differently, and, as a result, the automation will have different effects on users' workload (McBride et al., 2010). Additionally, variations among users' fine motor abilities should be considered as those users with lower fine motor abilities (e.g., lower arm-hand speed) may require longer reaction times (Jipp, 2012). Users' implicit attitudes as well as user moods and emotions may impact their trust in (Merritt et al., 2012) and reliance on (Merritt, 2011) automated systems. Furthermore, automation may be useful in helping those users with decreased functional states maintain high levels of performance, albeit as the expense of secondary task performance and an increased risk-of-return-to-manual performance decrements (Reichenback et al., 2011).

SECTION 5. AUTOMATED DRIVING-RELEVANT DATABASES

Several instances exist of real-world data collection using L2- and L3-type automated vehicles, many of which have been discussed within this document. In the interest of providing an initial reference for future explorations of such data, the following list of projects is provided. While this list is not meant to be comprehensive regarding all research examining automated vehicles, it is meant to specifically include those studies that have examined concepts relevant to near-term L2 and L3 automated vehicles for use on public roads. It is important to note that many manufacturer initiatives are not included here as they have not been made publicly available. Additionally, many technical and legal challenges may exist in trying to access these data. However, the following studies provide some evidence of the number and type of data collection efforts that have occurred to date.

Ardelt et al., 2012: In conjunction with the HAD project, data were collected over the course of a 65- km test drive.

Bertozzi et al., 2012: As part of the VisLab Intercontinental Autonomous Challenge, it is estimated that approximately 50 terabytes of data were collected. These data offer a resource for algorithm development. Researchers have undertaken efforts to rerun the trip in virtual space on an Italian supercomputer in order to evaluate algorithm performance. Their stated intention is to make the data set available to other researchers going forward.

Chan, 2012: The SARTRE project included a demonstration of a road train with three and four vehicles; several hundred km of testing were performed with the cars driving behind the lead vehicle at 85 km/h, separated by 5–15 meters.

Google: Google has collected over 300,000 mi of L2-type vehicle driving data.

Kessler & Etemad, 2012: The European Large-Scale FOTs on In-Vehicle Systems (euroFOT) resulted in the collection of data from vehicle sensors, video streams, and questionnaires. Protocols are in place to allow part of the data to be accessed by third parties. In total, almost 35 million km (approximately 22 million mi) of data were collected over 598 total hours of driving. After June 20, 2013, access to the euroFOT database was not defined in the description of work; therefore, access to the data will have to be negotiated separately with each of the five vehicle management centers.

Tsugawa, 2012: The EnergyITS project demonstrated three heavy (25-t) trucks at 80 km/h with a 10- m gap; in March 2013 a scheduled demonstration of three heavy trucks and one light truck at 80 km/h with a 4-m gap was held on 24-km (8 km – 3 times) of expressway before public use. About 100-km of test track (oval) data were also collected. Market research was conducted along an expressway (80 km/h, 10-m gap).

Zlocki, 2012: The KONVOI project gathered over 3,000 km of test data collected on closed test tracks, closed motorways under construction, and on motorway BAB 1 in real-world traffic. Configurations tested include a 2-truck KONVOI, 2 × 2 truck KONVOI, 3-truck KONVOI, and 4-truck KONVOI. Analysis included up to 2,400 overtaking maneuvers, 13 vehicles entering between the KONVOI trucks at junction, and 2 vehicles entering between KONVOI trucks at junction during coupling and decoupling maneuvers.

Flemisch et al., 2010: During the evaluation of two different transition designs for the HAVEit Temporary Autopilot, data were collected on a 70-km (approximately 43.5-mi) stretch of the A39 motorway. Vehicles traveled at a maximum speed of 100 km/h (approximately 65 mi/h) over an approximately 45-min time period.

Strauss et al., 2010: Validated the ARC system as part of the HAVEit project.

Schoenherr, 2009: As of 2009, the U.S. Army's Convoy Active Safety Technology System has collected more than 161 km (100 mi) of continuous autonomy with a speed of 80 km/h (50 mi/h).

Viti et al, 2008: During a large-scale FOT, 19 vehicles were driven in real traffic across the Netherlands for 6 months. During the first month, the ADAS (ACC and LDWS) were inactive to allow participant to get acquainted with the vehicles (Volkswagen Passats). For the following 5 months, participants were allowed, but not required, to use the ADAS. Data were collected on vehicle driving characteristics (e.g., speeds, accelerations, braking force, and times and positions on the roads), on the ACC usage (activity status of the ACC, speed limit, headway set), and on the preceding vehicle characteristics (distances, speeds, width). GPS positioning data were recorded from the full-traffic cars to map-match signals with the traffic flow database using the road-monitoring system Regiolab, which covers a number of motorways in the Netherlands.

SECTION 6. MANUFACTURER APPROACHES TO VEHICLE AUTOMATION

This section provides an overview of current industry approaches to vehicle automation. This discussion has been divided to include a summary of industry motivations, concepts associated with highway driving and highway platooning, approaches for city street driving, automated valet parking innovations, automated operations on unimproved roads, and a review of the limitations of enabling technologies. The section concludes with a review of the role of automated vehicle systems.

Industry Motivation

Representatives from various automakers have made public statements regarding their companies' individual motivations for pursuing automated vehicle production. A survey of these public statements was made and are depicted in Figure 6-1; brief summaries of each automaker's statements appear below.

Audi

Audi has defined their automated vehicle concept as piloted driving (Audi of America, 2013). These systems assume the driving task for a limited period of time, thus contributing to more comfort during the ride.

BMW Group

BMW Group (2013a) announced a new research and development partnership with Continental focused on automated driving. Their plan, which is motivated by road safety and a Vision Zero (Accident-Free Mobility) aim, calls for highly automated driving functions ready for implementation in the 2020 timeframe. These goals are supplemented by goals for improved convenience and efficiency. Partially automated driving functions of the near future, such as the Traffic Jam Assistant, are seen as an important step on the road to highly automated driving. Both BMW and Continental agree that highly automated driving will play a major part in ensuring future sustainable personal mobility.

In their observations, BMW Group (2011) described drivers and passengers experiencing highly automated driving. While initially reporting a "strange feeling," drivers and passengers begin to relax and trust the system after a few minutes. Nevertheless, with the BMW Group plan, the driver must stay aware of his or her surrounding conditions and remain responsible for the situation at all times.

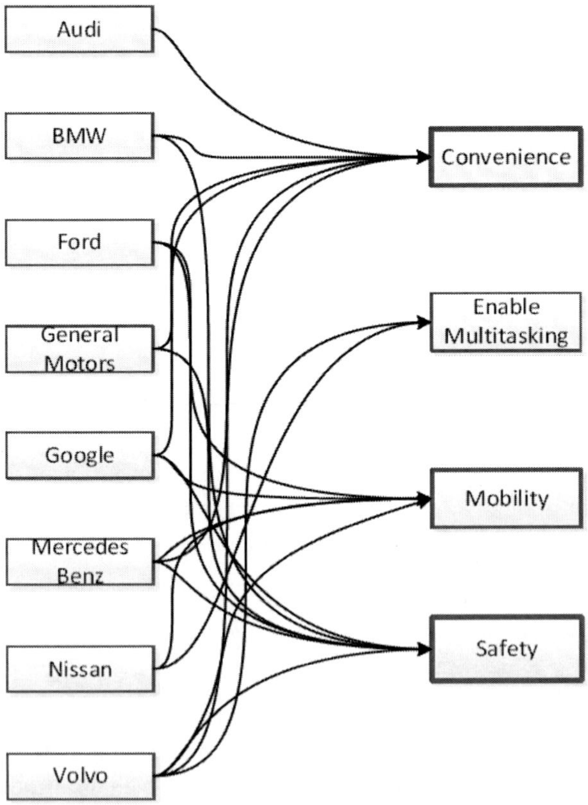

Figure 6-1. Industry Motivations for Automation (Source: Author).

Ford Motor Company

Ford Motor Company's (2012) *Blueprint for Mobility* outlined the company's views on the evolution of transportation in the near term and beyond. Based on forecasts showing that the 1 billion cars on the road today could double, or even quadruple, by mid-century, Ford's blueprint addresses the technologies, business models, and partnerships needed to address the challenges of population growth and urbanization. Ford plans to take advantage of the car as a rolling collection of sensors to reduce congestion and help prevent accidents. The development of these technologies is the first step in Ford's plan for a more connected future which will save time, conserve resources, lower emissions, and improve safety (Ford Motor Company, 2012). As such, Ford sees automation coming in stages, starting with freeway driving (Hachman, 2012).

Ford's emphasis on the driver's role is consistent with Nevada's requirement that a driver must be unimpaired and in the driver's seat (Hachman, 2012). To that end, Ford is working to establish a rapport between the car and its driver. This rapport includes the drivers' ability to quickly assume control in an emergency, as well as establishes driver confidence that the car can act independently and safely. While working towards opportunities for automated driving, Ford aims to keep the driver connected and involved in the decision-making loop. Such a system, if performing robustly and avoiding wrong decisions, will be one that consumers find trustworthy (Hachman, 2012).

General Motors Corporation

GM's overarching philosophy is that the technology should make driving safer and easier but not fully remove the driver from the equation (Bunkley, 2012). The driver should remain in the loop, engaged and aware of the situation, and ultimately responsible for the vehicle's operation. For example, the Super Cruise system keeps the driver in the loop through a series of driver alerts (GM, 2013). GM considers that the transition from automated mode to the driver taking control has to be done in a timeframe of approximately 2–3 s (Smyth, 2012). Going to fully automated vehicles, GM expects it will be possible to let people do other things in the vehicle, but they see this as much longer term (Smyth, 2012). The high prevalence of drivers talking on cellphones, texting, and doing other activities that take their attention off the road helps make a case for vehicles to take over in at least some situations (Bunkley, 2012). GM noted that the issues raised by self-driving vehicles are a key reason they have decided to take their Super Cruise, Cadillac's semi-automated driving system, to the next stage of development, including real-world driving assessment and trials (Bunkley, 2012; GM, 2013). Salinger (2013) provides an overview of GM's rollout sequencing, beginning with Driver Information and Alerts (no control; in production 2008), Emergency Intervention (limited control; in production 2013), and moving on to Limited On-Demand Automation (monitored control), Complex On-Demand Automation (transferred control), and Autonomous Driving (Chauffeured Driving) as future steps.

Google

Google indicated that they are developing automated vehicles as part of their goal to use and create technology that improves the world (Silicon Beat, 2012). Google noted that automation can be used to save lives, reduce the nation's energy consumption, free up substantial time every day (approximately 52 min per working American), triple the capacity of the highway system, and enable new models of car-sharing (Hansen, 2011). In practice, Google has been testing its cars on public roads since 2010, working with California, Nevada, and Florida to gain access to its roadways. Results of two studies analyzing Google's self-driving Prius and Lexus cars found the cars to have safer and smoother steering than when a human is in control (Simonite, 2013).

Mercedes Benz

Mercedes-Benz Research & Development North America, Inc., stated that automated driving is part of their vision for the future of their automobiles. Mercedes-Benz claims to see automation as providing greater safety, relieving the driver of cumbersome routines and/or tedious situations (customer convenience), and enabling the safe use of entertainment, telematics, and communications systems (enabling multitasking). Societal benefits associated with these efforts include the capability of enabling the elderly and disabled to travel with their own vehicles plus the increases in fuel economy through better traffic flow (Association for Unmanned Vehicle Systems International [AUVSI], 2012).

Nissan Motor Company

Nissan Motor Company's basis for autonomy is safety first (Sierhuis, 2013). The company philosophy in this domain is to provide more safety, i.e., "our cars never cause accidents." A distinction is drawn between "driver initiated autonomous driving" in which the vehicle monitors for threats continuously and drives autonomously when it intervenes to handle a threat, and "car initiated autonomous driving" which activates when the car drives better than the driver. An important concept introduced here is that of the car and driver working as a team, akin to the partnership of equestrians and their horses. Nissan has also noted that an aging population with serious purchasing power and a hunger to retain their mobility as they get older will likely be the reason why driverless, or less-driven cars, are adopted into the mainstream (Bigman, 2013). Additionally, Nissan noted the immediate benefits of automation to older people and those with handicaps. Further, people who cannot afford cars will be able to access car-based mobility due to car sharing. From a broader viewpoint, society, urban design, and transportation as a whole may change as individuals will now be able to undertake new activities in the car which were previously impossible while driving (Nissan Motor Company, 2013).

Toyota Motor Corporation

Toyota Motor Corporation (2013b) reviewed the company's Integrated Safety Management Concept –which includes active safety systems – and noted that components of these research efforts could lead in the future to a fully automated car. However, the vision is not necessarily a car that drives itself. Instead, Toyota and Lexus are focusing on technologies that support the driver, with the philosophy that a more skillful driver is a safer driver. Toyota noted that as automated efforts move forward, the driver needs to remain fully engaged in the driving process, acting as an intelligent, always-attentive co-pilot (Toyota Motor Corporation, 2013b).

Volvo Car Corporation

The vision for Volvo Car Corporation "is that no one is killed or injured in a new Volvo by 2020" (Duxbury & Stoll, 2012, para. 1). Automated driving is important to this vision of a future without car crashes (Volvo Car Corporation, 2012). Coelingh (2013), observing that in over 90% of all crashes human error is partly or fully responsible, notes that eliminating human error offers the largest potential in reaching their safety target; further, an automation system that does not require supervision can be made safer than one that relies on human supervision.

Anders Eugensson, Volvo's head of government affairs, noted that the vehicle integration of wireless Internet would result in the interconnectivity necessary for the coordinated operations of road trains. These road trains would operate in a manner similar to a team of horses (Duxbury & Stoll, 2012):

> The car of the future will be just like the farmer's horse. The farmer can steer the horse and carriage but if he falls asleep the horse can still take him back home. And if the farmer tries to steer the carriage against a tree or off a cliff, the horse will refuse.

Automated driving is not only seen as safer, but it is also seen as a way to increase driver productivity. Volvo Car Corporation (2012) asserted that automated driving paves the way for more freedom behind the wheel, creating the possibility to safely do something else, such as sending text messages or reading a book while the car is driven in an automated mode. The first focus areas in Volvo's technology development are automated driving in slow-moving queues and, with a longer perspective, road trains on motorways. One of their research conclusions is that younger consumers, in particular, are willing to pay for technology that can help manage the distractions created by the urge to be constantly connected, even when traveling in a car. Volvo Cars sees highly automated driving entering the market gradually (Coelingh, 2013).

Highway Driving

Traffic Jam Assist

Several automakers have announced various implementations of traffic jam assist (TJA). These systems are either already present on model year 2014 vehicles, or are present on forthcoming products. For these automation concepts, TJA may be broadly viewed as operating in an L2 modality.

Several production system announcements for model year 2014 vehicles are of note. Mercedes has announced the availability of TJA (i.e., Distronic Plus® with Steering Assist and PRE-SAFE Brake features) on their S-Class vehicle for model year 2014 (Mercedes-Benz USA, 2014); Daimler AG, 2013, 2012a). The Distronic Plus functions within a speed range of 0 to 200 km/h (approximately 0 to 124 mi/h). The Mercedes system employs a sensor suite enhanced with stereo vision, which enables normal-lane marker detection as well as car-following when in a traffic jam. Speed is adapted to the flow of traffic ahead, slowing as necessary. The system can brake to a full stop. When traffic moves, the driver can resume with a tap, or, if the stop is less than 1s, automatically (Mercedes-Benz USA, 2014). The driver is expected to be engaged and have hands on the steering wheel. If the driver is not touching the steering wheel, the system issues a warning and disables after a few seconds. Volvo Cars has announced plans to launch TJA in 2014, for initial use in heavy traffic at speeds up to approximately 59 km/h (31 mi/h; Duxbury & Stoll, 2012).

BMW Group (2013b) notes that TJA will be available on the 5 Series vehicle starting in November 2013. The TJA system in the BMW i3 Concept is described in BMW Group (2011). The TJA function helps drivers in monotonous traffic situations and congested areas, taking over vehicle control to reduce driver stress levels. TJA maintains a safe distance between vehicles and automatically controls the speed – down to a standstill and restart – with the ability to bring the car to a complete stop. The system provides active steering support as well; however, the driver is required to keep one hand on the steering wheel. TJA operates at speeds up to 40 km/h (approximately 25 mi/h).

Audi's *Piloted Driving* was demonstrated at the 2013 Consumer Electronics Show on a Nevada freeway in public traffic (Audi of America, 2013). The aim of Audi's system is to reduce the driver's workload in stressful situations, such as in congested traffic. The system provides automated control of steering, braking, and throttle at speeds of 0–60 km/h and builds upon S&G ACC. Sensing includes radar (250-meter range with a 35-degree field of view), a wide-angle video camera, a laser scanner (80-meter range with a 140-degree field of view), and eight ultrasonic sensors monitoring zones directly in front of the car and at the corners. The system can detect objects such as pedestrians, other vehicles, and guardrails. Perception sensors enable the vehicle to define a *de facto* lane in the absence of lane markings. The system also responds cooperatively to cars moving into or out of the lane. The Audi TJA system monitors the subject vehicle's speed and nearby traffic. If it detects traffic jam conditions at speeds below 60 km/h, the TJA function enables and the driver can activate the assistance function. If the vehicle reaches the limits of the function, such as dispersal of the traffic jam, the driver is prompted to take over control.

According to Ford Motor Company (2012a) the company is developing TJA with an anticipated mid- term (2017 to 2025) implementation. The system will use radar and camera technologies to automatically keep pace with other vehicles and provide automated steering control to stay in the current lane, thus reducing driver stress and potentially improving vehicle flow. Ford noted that their system employs technology from features already available in Ford cars, including Active Park Assist, ACC, Lane Keeping Aid (a lane departure avoidance system), and Ford Powershift transmission. The press release noted that Ford's TJA system is intended for environments in which there are no pedestrians, cyclists, or animals, and where lanes are clearly marked. Additionally, Ford's TJA system will be able to communicate any developments regarding surrounding traffic to the driver. In doing so, the system incorporates features to help the driver remain alert and in contact with the vehicle controls, even when the system is active. Drivers will also be able to override the system at any time.

Highway-Speed Automation

Level 1 Systems

Nissan Motor Company (2013) announced a new capability in model year 2014 Infiniti vehicles. These vehicles will be equipped with steer by wire, which allows for the independent control of a vehicle's tire angle and steering inputs. This capability insulates the vehicle from unnecessary road-generated disturbances such as ruts or crosswinds. Steer by wire enables a new straight-line stability system which assists on-center driving. The system reduces discrepancies between the intended path and the actual path; in addition, the company asserts that the driver will be required to make fewer fine- grained steering adjustments, which will reduce driver fatigue and workload.

L2 Systems

Several L2 systems intended for highway operation have been discussed in the community. GM's *Super Cruise* system provides full-speed-range ACC and lane centering, and uses cameras and radar to automatically steer or brake in highway driving (GM, 2012). GM's Super Cruise system is based on an integrated combination of radar, ultrasonic sensors, cameras, and GPS map data (GM, 2013). This technology results in a system capable of semi-automated driving, including hands-off lane following, braking, and speed control under certain driving conditions (GM, 2013). The system allows the driver to take his/her hands off the steering wheel for extended periods, and is intended for operation in highway driving scenarios. GM safety engineers note that drivers may engage in secondary tasks while using Super Cruise, and work is underway to develop techniques to appropriately manage secondary task behavior (GM, 2013). Noted benefits of the Super Cruise system include reduced driver workload in freeway conditions, bumper-to-bumper traffic, and on long road trips. While not yet available, Super Cruise–enabling technologies, such as 360-degree crash-risk detection and enhanced driver-assist features, are available on model year 2013 vehicles as part of GM's Driver Assist Package.

In Europe, the Honda Accord has been available with LKAS since 2008. The LKAS provides up to 80% of steering force required, with the driver expected to provide the remaining 20% (Honda Motor Europe Ltd., 2008). These vehicles can also be equipped with ACC, making simultaneous LKAS/ACC possible. The system operates at between 72 and 180 km/h. Honda noted that the system is designed to complement the driver, not to replace his/her input; therefore, a Honda Accord with the combined LKAS/ACC system could be considered an L2 vehicle under current NHTSA definitions. Price (personal communication, April 27, 2009) further noted that a warning is provided if the system detects that the driver has taken his/her hands off the steering wheel.

BMW Group (2011) described demonstrations of an L2 vehicle which provides lateral and longitudinal control at highway speeds. When slower traffic ahead prevents the vehicle from maintaining its set speed, the vehicle monitors the adjacent lane and performs a lane change when safe. The vehicle also adjusts speed to allow merging vehicles to enter traffic safely. The system can function up to a speed of 130 km/h. Sensing is performed by a fusion of lidar, radar, ultrasonic, and camera sensing technologies. In 2011, a BMW test vehicle was tested in public traffic, traveling on the A9 motorway between Munich and Nuremberg with no driver intervention (Ardelt et al., 2012). During 2013–2014, BMW will be working with Continental to equip several near-production L2 vehicles, which will be driven by a selected group of trained test participants in pilot testing. This testing will be conducted on German and European motorways and will take into account motorway intersections, toll stations, road work, national borders, and other challenging situations (BMW Group, 2013a).

Toyota Motor Corporation (2013a) described its advanced safety research vehicle at the 2013 Consumer Electronics Show. The vehicle includes an LKAS with yaw rate feedback control. When differences between ideal and actual vehicle posture caused by uneven road crown (camber) or crosswinds are detected, steering torque is automatically adjusted according to vehicle speed, thus assisting the driver in staying in a chosen lane. The system operates in two modes. When ACC is inactive, the system offers an LDWS function only, including momentary corrective steering. If LKAS is activated while ACC is in operation, the system offers lane-keeping functions at speeds of approximately 72 to 177 km/h (45 to 110 mi/h).

Bosch's roadmap for fully automated driving addresses two regimes spanning multiple automation levels: a) system permanently supervised by a driver, and b) reduced driver supervision of system (Becker, 2013). The first regime includes offering combined ACC and lane keeping as well as Integrated Cruise Assist (ICA). ICA offers automated longitudinal and lateral guidance from 0 to 130 km/h (on highways and major roads; 0 to approximately 81 mi/h); if lane markings are inadequate, lateral guidance is provided via dynamic radar objects. ICA extensions (with driver confirmation) are automatic lane changes and automatic speed adaptation based on road sign recognition. Testing is underway on public roads in Germany and the United States (California and Michigan).

The supplier Valeo's efforts include their Cruising Assistance development program (Mattern, 2013). Within the Valeo system, the performance envelope is described as forward, and lane-change movements are made at speeds up to 120 km/h with a 120-m sensing range. Relevant objects are moving cars (same direction), road borders, lane markings, and speed signs.

L3 Systems

Google has developed an automated vehicle that currently can operate up to 75 mi/h at L2 in highway use and L3 for testing purposes (e.g., research being conducted in a safe setting). The vehicle does not change lanes. It will slow down/stop for moving traffic in front of it. The vehicle combines laser, radar, camera, and map data to stay in its lane in an automated driving mode until the user is ready to retake control. An in-dash user interface displays system-state information and is used to notify the driver when automated mode is available, as well as to prompt the driver to retake control when needed.

Moving from L2 to L3 systems has significant implications in system design (Schumacher, 2013). Describing supplier Continental's Vision of Accident-Free Driving (Vision Zero), Schumacher noted, "We are convinced that innovative technologies will make it possible to one day drive a car without any accidents – in all vehicle categories and markets of this world" (2013, p. 2). Schumacher (2013) further stated that the shift from L2 to L3 requires a corresponding shift in the fail-safe concept: the driver's instinct and knowledge are no longer in the control loop, and the driver cannot serve as a fallback controller. The system must act as the fallback, which requires redundancy in signals, software, and electrical systems, particularly focused on braking and steering.

Within the Bosch automation roadmap, prototypes have been developed which require steadily decreasing levels of driver supervision (Becker, 2013). A highly automated "highway pilot" operates without the driver in the loop, and overtaking maneuvers are performed automatically. The Bosch system also reacts to construction zones and intersections. A further evolution has been defined as a fully automated "commuting" system. Testing is underway on public roads in Germany and the United States (California and Michigan). In the event of a system failure, the Bosch approach calls for the vehicle to be brought to a safe state – even without driver intervention; this effort requires redundant actuation, an emergency operation mode, and energy backup.

Emergency Automated Operation

BMW Group (2013a) described an ESA prototype that does not require driver involvement. If biosensors detect a medical emergency in the vehicle (such as a heart attack), the ESA is able to bring the vehicle safely to a stop. The system activates the hazard warning lights and maneuvers through traffic in a safe and controlled way to come to a stop on the road shoulder. An emergency call is automatically sent out as well to request medical assistance and notify the traffic authorities.

City Street Driving

The majority of published research and development work examining city driving has been undertaken in the European community. Efforts from multiple universities and consortiums in Europe have sought to develop automated driving concepts suitable for city street use. Additionally, a recent DARPA Grand Challenge awarded a prize for automated operations on urban roads. These efforts are discussed within this section, along with supplier activities.

The 2007 DARPA Urban Challenge attracted dozens of teams with unmanned automated vehicles which could operate among each other in a residential street setting (DARPA, 2013). The top three finishers (CMU/GM, Stanford, and Virginia Tech, respectively) successfully handled traffic circles, four- way stops, parking lots, and road obstacles on these streets.

The University of Braunschweig in Germany has built upon their entry into the DARPA Urban Challenge with their Leonie vehicle (Nothdurft, 2011). Their research focuses on automated urban driving, and they regularly conduct testing in live traffic on the city's *Stadtring*. This busy 4+ lane arterial has such features as traffic lights, turn lanes, and pedestrians. The vehicle executes turn maneuvers, responds to traffic lights, and performs lane changes as needed to flow in traffic.

Mattern (2013) describes supplier Valeo's Urban Assistance development program. The performance envelope is described as forward movement and 90-degree turns, speeds up to 50 km/h (approximately 31 mi/h), and sensing range of 50 m (approximately 55 yd). Relevant objects are cars at all speeds (moving and parked), pedestrians, infrastructure, road signs (yield), and traffic lights.

The German UR:BAN project is also aimed at extensive situational awareness in urban environments (UR:BAN, 2013). While the project is not explicitly aimed at automation, safe lateral and longitudinal vehicle control are two of its stated areas of interest. Using panoramic sensing, systems and algorithms are being developed for driver support in driving on narrow or obstructed streets, resolving conflicts with opposing traffic, and performing lane changes.

The European CyberCars project developed urban vehicles for operation in city centers in lanes or districts segregated from normal traffic (Vlacic, 2009). Vehicles were capable of performing driving maneuvers in cooperation with each other, including overtaking, platooning, docking and undocking, roundabout driving, intersection crossing, obstacle avoidance, and path generation.

Additionally, a specific V2V and V2I communication architecture was designed, developed, and implemented to enable interconnectivity and interoperability between the vehicles. Public demonstrations were held in eight cities in Europe. INDUCT (2013) described NAVIA, a driverless eight-passenger robotized shuttle designed for transportation in such areas as city centers, pedestrian areas, private campuses, and airport parking lots. NAVIA is equipped with laser range finders, cameras, and a software package enabling safe automated operation in any environment, according to the company's website. At the 2013 Driverless Car Summit, Lefevre (2013) provided additional details about the system. As currently designed, the maximum speed is 12.5 mi/h (20 km/h). Four- wheel steering enables high maneuverability in tight urban spaces: the turning radius is less than 3 m. Battery charging is inductive. The perception system consists of four laser scanners for 360-degree coverage, two stereo fusion cameras, eight ultrasound sensors, plus GPS. Combining this data with odometry, the system is able to recognize traffic signals and obstacles, identifying obstacles as pedestrians, cars, or bicycles.

Automated Valet Parking

Many suppliers and automakers are working on or have developed automated valet parking solutions. These differ from currently available parking assistants (such as the self-parking systems currently available from Toyota, Ford, and Volvo) as the driver is not expected to be in the vehicle while automated valet parking is occurring. While this automation concept does not fall under the on- road/driver present scenarios described by NHTSA automation definitions, they can be seen as a foundational automation technology for vehicles. Further development of automated parking is likely in the near future.

Audi of America (2013) has developed a prototype system which enables the driver to depart the vehicle at the entrance to a parking garage and instruct it to proceed to park itself via a smartphone interface. The vehicle then travels without a driver into the garage and parks in an empty space. Upon being summoned by the owner via smartphone, the vehicle proceeds to the garage exit area for the driver to re-enter the vehicle and resume driving. The parking facility's central computer takes over part of the control function and guides the vehicle via wireless local area networking (WLAN) to the nearest available parking space. The vehicle's movements are monitored via external laser sensors. The vehicle also uses 12 ultrasound sensors for situational awareness; in the future, four video cameras will also be used. Audi is currently equipping a parking garage in Germany for further testing of this system.

Williams (2012) reported on Nissan's self-parking car. The system enables a driver to exit the vehicle at the entrance to a parking area and tap a *park in* button on a smartphone car application; the car then travels without a driver and parks. Nissan notes that, upon receiving the park request, data is sent to the Nissan Global Data Center, which performs a vehicle health check. If vehicle health is confirmed, the automatic driving mode is enabled. The vehicle relies on a digital map of the parking area supplied by a cellular data link. For maneuvering and localization, four cameras are used. The article quoted Nissan as saying this is a more accurate method than using GPS. At a trade show demonstration, a car emblazoned with a large 2015 on its side drove at about 5 km/h while in automatic driving mode.

Cunningham (2011) described a system developed by Tier 1 supplier Valeo which was demonstrated at the 2011 International Motor Show in Frankfurt. The vehicle's parking maneuver was initiated by a command from an iPhone. Perpendicular parking was accomplished via automatic control of steering, brake, and throttle. The article noted that these types of systems would be useful for parking a car in narrow spaces, where it would be difficult to open the driver door once parked, and for wheelchair unloading, which might require more space next to the car than is available. Mattern (2013) noted that this system followed development of automated steering (2007) and automated braking (2010), with automated throttle, automated gearbox, and remote control now in the product development phase. Termed Maneuvering Assistance, the system performance envelope is described as omnidirectional movement under 6 km/h (approximately 4 mi/h), with a 6-m (approximately 7-yd) sensing range. Objects to be detected are mostly stationary and include parked cars, trees, curbs, and pedestrians.

Enabling Technologies: State of the Art

The technologies to accomplish automated driving fall into the categories of situational awareness, decision-making, and actuation. Technology for situational awareness includes sensors, maps, positioning, and sensor fusion. Decision-making is software-based, requiring highly sophisticated algorithms at the planning and control levels. Actuation requires robust and fail-safe control technologies. Testing and certification can be seen as another enabling technology area. Various reports and industry information provide some insight into the state-of-the-art and technical challenges as seen by the practitioners.

Perception

Broggi (2012) noted that since the DARPA Challenges work in perception has split into two main areas. The first includes high-performance perception using currently expensive lidar-based sensors which require challenges in vehicle integration. The second area focuses on low-cost and integrated stereo-vision sensing, which is also aligned with military and off-road commercial applications such as mining and agriculture. Current high-performance lidars generate 2.5-million distance estimates per second but with deformations in the resulting data when the sensor is mounted on a moving platform, while stereo vision experiences higher noise than lidar but still can generate about 1.7-million correct measurements per second. Additionally, lidar systems are more sensitive to smoke, rain, and dust than is stereo vision. Broggi (2012) further noted that that image data can be useful in providing texture data, which can in turn assist in pattern recognition. He concluded with the observation that those investing in lidar-based systems are seeking to lower costs, while those investing in stereo- vision systems seek to increase performance.

AUVSI (2012) reviews comments made at the 2012 Driverless Cars Summit. Invited experts at the event discussed challenging technology issues. Unusual obstacles, which are rare but must be addressed, were described as a Mylar balloon (with a large radar cross-section) or a bouncing ball (likely to be followed by a child racing to retrieve it). The complex scenarios on city streets involving pedestrians, bicyclists, unstructured intersections, etc., were cited as further examples of the challenges being faced. In particular, pedestrian interactions with cars frequently involve eye contact between the pedestrian and the driver to confirm they are aware of each other. The MIT MediaLab's City Car for urban driving aims to develop automated vehicles capable of sensing other people and objects and reacting intuitively. An example of this intuitive behavior would be the recognition of a pedestrian with headlight movements that mimic eye contact (AUVSI, 2012).

A Google representative, when asked about situations the Google car cannot handle, cited erratic behavior by other drivers or extreme and rare situations such as a vehicle entering the highway the wrong way via an off ramp (AUVSI, 2012, p. 7). Additionally, Muller (2013) discussed challenges with Chris Urmson of Google's self-driving car team. Google vehicles rely on a database created by humanly driving the road; this data is added to highly detailed maps of the roads and terrain. When the automated system is driving the vehicle, it is comparing real-time sensor data to the database for greater robustness in perception. However, Urmson noted that due to the dynamic nature of the environment, the Google system cannot perform adequately in heavy rain or on snow-covered roads (Muller, 2013, para. 10). Additional challenges exist when the vehicle encounters a rare event, such as a stalled vehicle, or road debris (e.g., a tire carcass) (Muller, 2013). Urmson (2012, para. 3) also noted the challenge of interpreting signs and signals at temporary road construction zones.

Rodriguez (2012) discussed the challenge of detecting animals on the road at night and described a new system developed by Autoliv which uses data from thermal imagery infrared cameras to detect animals that are beyond the vehicle headlights' field of vision. The technology is the next generation of a night vision product that is currently installed on several car models. Autoliv was quoted as saying they have contracts with several OEMs to offer the animal detection system as an option as early as 2013.

Daimler (2012b) described a new perception technology they developed called 6D Vision which identifies hazards in complex traffic environments. Based on stereo camera data, algorithms resolve both the position and the movement of the three dimensions of objects; hence the name 6D Vision. Daimler noted that the system will go into production in 2013.

A cyclist in the same lane swerving out in front of the car is one incident type occurring in urban driving. Volvo Car Group (2013) announced a new system to detect and automatically brake for cyclists in this scenario, as an enhancement of present pedestrian detection and auto brake technology. All cars previously equipped with pedestrian detection will now also incorporate cyclist detection. The system relies on combined radar-camera data. The radar has a wide field of view, and the high-resolution camera allows for robust detection in the complex urban traffic environment.

Following the SARTRE project, Volvo Cars concluded that their production sensor system is "partly sufficient" for road train operation; basic steering and brake control is sufficient (Coelingh, 2013). While lateral control within the platoon was described as "difficult," the cost of supporting communications was considered to be affordable. The main challenges for automated driving – dependability, verification, and sensing – were described as follows (Coelingh, 2013):

- Vehicles need to handle at least single faults; systems should be "fail-operational."
- ISO 26262 is not sufficient for verification.
- Improved sensor detection capability is needed to minimize false negatives.
- In terms of actuators and power, redundancy may be needed in case of failure.
- Verification of very low-frequent events is needed.

Salinger (2013) described the increased perception performance needed as increasing levels of automation. Three broad areas of requirements are identified: navigation and active safety requirements, on-demand assistance and crash avoidance requirements, and driverless operation requirements. Numerous challenges for automation driving identified pertain to sensing and perception capabilities, fault-tolerant/fail-safe vehicle control, strong situational analysis abilities, appropriate system responses for emergency situations and rare-events, interactions with non- automated vehicles, driver inattentiveness detection, positioning challenges (e.g., GPS lane level accuracy and availability, digital map accuracy, virtualization), and security/privacy concerns. To address the challenges associated with automated driving, enabling technologies are needed in the following areas (Salinger, 2013):

- Sensors
  - Object sensing: smaller/easier to fit on the vehicle, less expensive, higher resolution (range, horizontal, and vertical angle), larger field of view (longer, wider), higher update rates and lower latency
  - Road sensing: sign/traffic signal information, lane geometry, surface friction
  - Driver state sensing: attention, intent
- Fail operational functionality
  - Sensing
  - Actuation
  - Processing
  - Communications
  - Power
- Networking and infrastructure information
  - Maps/GPS
  - Lane level information
  - Faster update rates
  - V2V and V2I communication

Bosch (Becker, 2013) noted that perception for high levels of automation requires "surround sensing" which entails 360° surround view, 3D information, shape and surface measurements, high reliability, low sensitivity to weather and lighting conditions, physical redundancy, active measurement (preferred), and object classification (preferred).

Similarly, The Karlsruhe Institute of Technology has identified the challenges of perception for driver assistance systems in urban traffic scenarios. This research, which uses a stereo camera approach, has identified the following unique challenges associated with urban traffic (Karlsruhe, 2013):

- Crowded situations with many types of traffic participants such as cars, pedestrians, cyclists, trams, etc.
- Objects that may (partly) occlude each other and may look considerably different from different viewpoints.
- Scenarios which may change abruptly compared to highway driving.
- Wide variation in road geometry.

- Features typically relied upon, such as lane markings, can be partly occluded or completely missing.

Additionally, Team AnnieWay has been working in the area of environmental perception for automation driving, focusing mainly on vision-sensing capabilities (Karlsruhe Institute of Technology, 2013). A Volkswagen Passat is equipped for fully automated driving and an Audi Q7 serves as mobile sensor platform and as communication partner in cooperative driving. Sensing technology consists of stereo cameras, radar, a Velodyne 3D lidar, GPS, an inertial measurement system, and communication devices for V2V communication. Current areas of investigation are:

- Efficient large-resolution stereo vision
- Optimal path and trajectory planning for autonomous vehicles
- Intersection detection and estimation from video sequences
- Scene understanding from 3D lidar data
- Collaborative driving strategies and longitudinal control of vehicle convoys

Machine Intelligence and Decision-Making

Keane (2013) reported on comments made at the 2013 SAE Government-Industry meeting regarding technology challenges. At the meeting, Anthony Levandowski of the Google self-driving car team described Google's biggest challenge as ensuring the reliability of the decision-making software which controls the vehicle. Designing the Google system with proper processes to understand and minimize failure is essential. The resulting system needs to know how to respond to a variety of scenarios that may occur with little to no room for errors. Additionally, the system will need to be able to make judgments to understand situational contexts. For example, a ball rolling into the street may be followed by a child trying to retrieve it (Keane, 2013, Foolproof Software section, para. 3 & 4).

Coelingh (2013) also described decision-making and control as key challenges for automation, particularly in terms of robust sensing, vehicle positioning, situational awareness, control performance, and user experience.

Sierhuis (2013) introduced the concept of a "social car" acting in concert with the driver as a "team member," noting that human activity is always socially inspired. As people and cars interact, the question becomes: "What services does the car provide to the human and vice-versa?" Overall, the car needs to understand the daily life of the human team member. Sierhuis described key factors as coordination (sharing information), cooperation (aware of and supporting each other's goals), and collaboration (working on a shared project).

Karlsruhe Institute of Technology's (2013) Cognitive Automobiles team described investigations into techniques "to give automobiles an abstract understanding of traffic situations and to make autonomous behavior decisions on this basis" which draw upon models of human cognition. A cognitive automobile is described as one which perceives itself and its environment with sensors plus communication to gain information from other traffic participants. The team is developing a consistent probabilistic model of the vehicle's knowledge about its environment and itself; such a model is necessary since a vehicle can only partially observe its surroundings and these observations can be noisy. The researchers asserted that the vehicle must be aware of these uncertainties, with this knowledge flowing into the behavior decision process. An additional focus is in developing machine learning methods which incorporate automatic reasoning about the behaviors of other traffic participants and the risk potential of situations to implement anticipatory and self-aware decision- making. While a typical approach to automated driving behavior planning is to manually model decisions for different environmental states, their approach is to make decisions automatically for every probability distribution over the state space. This method enables the vehicle to explicitly consider its own uncertainty and lack of knowledge due to incomplete perception. A key aspect of this research is the integration of empirical knowledge the car gains in real or simulated driving experiments, i.e., gaining knowledge from a human "driving instructor."

Rupp and King (2010) discussed machine learning for situational awareness and decision-making. Noting that humans will never be perfect, the authors noted that humans are permitted to operate vehicles, and society accepts the consequences. Rupp and King wondered how much better a machine has to be than the human it is to replace before the replacement is societally accepted. They posited that the key to acceptance could be in the ability to learn; possible machine replacement of human drivers will come when the system demonstrates its ability to perform and learns at least as well a human and is able to avoid making the same mistake twice. An even higher level of performance could be a system which is designed to learn from and avoid the mistakes of other automated systems on the road. They recommended that future systems for situational awareness should be conceived with these limitations in mind (Rupp & King, 2010).

Rupp and King proposed that the next steps for situational awareness focus on improving the true versus false detection rate for key objects and events (2010). Accuracies of the target characteristics have to be interpreted, and the validity of the detection itself must be verified. The artificial intelligence must handle hacker attacks, unknown objects (such as a new type of vehicle that is not communicating), and internal system failures. Rupp and King see V2V and V2I network communications as an additional sensor that will improve the accuracy and timeliness of information when fused with other onboard sensors, extending sensing beyond line of sight and providing information about traffic, weather, road condition, and the states of nearby vehicles.

Rupp and King (2010) noted that the decision-making process followed by an automated vehicle system should act in a manner consistent with safe driving for the roadway conditions, as well as according to the general expectancies of other road users (e.g., the appropriate timing of lane changes). Part of the solution, as they see it, is for the system to learn the driver's specific driving technique and style; however, this ability is beyond the capabilities of current decision-making systems and suggests a variety of challenging research problems.

Test Methods

Several have commented on the need for robust safety testing. For example, Mercedes-Benz has noted that they expect new driver assistance systems to have completed millions of kilometers of error-free test driving before they consider it ready for mass production (Wüst, 2013). Urmson (2012) noted that Google vehicles have completed more than 300,000 miles of testing across a wide range of traffic conditions. To date, no crashes have occurred under computer control. Regardless, Silicon Beat (2012) noted that although Google's vehicles have traveled as far as 50,000 miles without a safety- critical intervention, Google strives to improve these numbers as the self-driving car will face greater scrutiny than a human driver would. Similar sentiments were echoed by others. At a recent meeting of the European iMobility Forum Working Group on Vehicle Automation, Bartels (2013, March) observed that specific development and test procedures are needed for testing highly/fully automated driving.

User Aspects of Automated Driving

Consumer Attitudes towards Automated Driving

Accenture (2011) described the results of a survey of more than 2,000 consumers which examined attitudes toward intelligent devices that frequently crash or freeze. About half of the respondents (49%) said "they would be comfortable using a driverless car" (2011, Keep It Simple section, para. 3).

J. D. Power and Associates (2012) reported the results of their 2012 Emerging Technologies Study. The study included questions regarding consumer sentiments on automated driving. Results were based on responses from more than 17,400 vehicle owners. Twenty percent of all vehicle owners say they "definitely would" or "probably would" purchase "autonomous driving" in their next vehicle after learning the estimated market price of $3,000. Prior to learning the price, consumers expressed only a 37% interest in this technology. After learning the price, premium vehicle owners maintained their interest at a level of 31% while the interest of non-premium vehicle owners was at 18%. The study found that vehicle owners were nearly as likely to select a fully automated driving mode as they were to select semi-automated driving technologies such as emergency stop assist, traffic jam assist, or speed limit assist. Vehicle owners with the highest interest in fully automated driving at market price were males (25%), those between the ages of 18 and 37 (30%), and those living in urban areas (30%). Those who expressed an interest in the automating parallel parking feature also indicated a preference in fully automated driving (41%); this finding illustrates similar consumer importance placed on semi-automated and fully automated driving modes.

Based on experiences and customer clinics in the SARTRE project, Volvo Cars concluded that autonomous driving is easy to adapt to "for most people" (Coelingh, 2013). However, selecting appropriate use cases and operating situations is critical; use cases such as joining the platoon from the side (rather than rear) and allowing non-platoon vehicles temporarily into the platoon caused "major problems." More generally, Volvo concluded that half of today's car buyers are ready to embrace autonomous driving and that "a majority of tomorrow's car owners will not buy a car without it."

To assess public perspectives on automated driving, Nåbo, Anund, Fors, and Karlsson (2013) described results from focus groups involving 28 participants who were asked to consider the future (5–20 years) and the possibilities that automated driving might bring. After an initial discussion focusing on general aspects of automated driving, the groups were shown automated driving video clips. Participant perspectives were captured during discussions and through a questionnaire. Concerns about affordability, legal issues, and loss of driving skill were voiced. The groups discussed various aspects of safety and security but were not clear if the systems would increase or decrease safety. Interestingly, participants were more supportive of systems in which the driver supervises automated driving than of systems in which the driver has a lesser role and can attend to something else. Additionally, some participants wanted to have automation during long, boring drives (i.e., for increased comfort), while others wanted it to help cope with difficult driving situations (i.e., for increased safety).

Rupp and King (2010) provided a perspective on the concept of a driverless car. Rupp and King asked (2010, p. 6): "Do consumers want a car without a driver, for instance to pick up the kids after school with no one in control on board?" Their contention is that taxi services, carpools, etc., already exist for that and other errands such as package pickup. Instead, consumers want a car that drives itself, relieving the owner of the immediate task of driving while retaining command. Rupp and King (2010) also hold that the driver should be in-the-loop, particularly during the transitional period in which vehicle automation is being proven within a complex traffic environment consisting of both machine- driven and human-driven vehicles. They drew the analogy of the driver acting in a supervisory mode, "like the orchestra conductor who commands all the instrumentalists (stop/start, faster/slower, louder/softer), but does not play the instruments himself" (Rupp & King, 2010, p. 13). This analogy implied that, even though the driver is not operating the controls, he or she must still maintain situational awareness. Their concept of *Full Driver* Assist provides the driver additional time, resulting in "more confidence in performing a more appropriate role in the overall system, one that is partially tactical but becomes mostly strategic in nature."

Rupp and King (2010) also noted that specialized training may become the future norm, such that the first automated systems can only be operated by drivers who have undergone specialized training, earning a certification and a special license to operate an automated vehicle. Interestingly, they also noted that, as automated vehicle technology becomes the norm, specialized and intense training may be required to operate a vehicle in *manual* mode (Rupp & King, 2010).

Driver-Vehicle Interactions

The DVI is an important element of an automated driving system. Sources provide some indication of drivers' propensity towards secondary tasks when using active safety systems applicable to automated vehicles. A key topic of discussion is driver monitoring and adaptive DVI.

Adaptive DVI

Benmimoun et al. (2012) reported on driver interactions with an ACC/FCW system within the euroFOT project, which could have some bearing on automated driving. Field testing found that drivers of passenger cars using ACC/FCW were three times more likely to engage in visual secondary tasks during normal driving. In testing with a separate fleet of vehicles, drivers were also more likely to engage in secondary tasks while driving with LDWS. However, in both cases this difference was not found during incidents. The authors concluded that this implies that drivers may be capable of managing secondary tasks and yet focus on the road ahead when the traffic situation requires doing so. In addition, ACC/FCW does not seem to affect the amount of drowsy driving. For trucks, no particular side effects on driver behavior were observed.

The German UR:BAN project is aimed at extensive situational awareness in urban environments (UR:BAN, 2013). The authors noted that novel assistance functions provide the driver with supporting information in complex traffic situations, but benefits arise only if the information flow is intelligently filtered to avoid overloading the driver. To this end, support is adapted to the driver's current state and activation by detection of overburdened or inattentive drivers. By incorporating adaptive support into the design of vehicle controls and displays, it is hypothesized that the driver will receive information much earlier and will be motivated to anticipate traffic situations rather than simply react to them. BMW Group (2012) noted that the UR:BAN researchers were developing methods of identifying the intentions of both the driver of the subject vehicle plus other drivers as early as possible, so that this information can be taken into account in the subject vehicle's response.

Salinger (2013) described a human factors study conducted by GM, based on the premise that the first autonomous systems introduced to the driving public may follow an "incorrect path." The project defined a Limited-Ability Autonomous Driving Systems (LAADS) as one which can control vehicle speed and steering on public roads for substantial distances and times while in some situations requiring that the driver/operator intervene to assure a safe and comfortable trip. The aim of the study was to investigate driver interactions with a LAADS to determine the effects of the system on driver visual attention to the driving task, driver willingness to engage in secondary non-driving-related tasks, and driver ability to respond to events. Further goals were to understand, in terms of human-machine strategies and control transition strategies, factors impacting the effectiveness of alternative concepts of operation. These issues were identified based on surveys, expert panel studies, driving simulator studies, and test track studies. It was found that riskier tasks (relative risk values greater than 1) tended to be limited to LAADS driving as compared to normal ACC driving. For instance, behaviors such as reaching into the rear compartment, extended glances to watch a DVD, phone interactions, and texting/emailing were significantly higher in LAADS driving. Subsequent work developed and tested countermeasures. Without countermeasures, off-road glance durations of up to 12 s were observed, whereas the maximum off-road glance time with countermeasures was about 4 s. By comparison, off-road glance times observed with ACC-only were under 3 s. The study concluded that driver engagement in secondary tasks is likely to increase when the level of automation provides the opportunity to do so; therefore, LAADS systems should be designed to clearly indicate the mode of operation and to encourage drivers to attend to forward roadway conditions. The study further concluded that HMI components can improve driver engagement, including providing:

- Means to engage driver in driving task when system is engaged
- Means to encourage visual attention to forward roadway
- Active alerts for system failures and limitations

Coelingh (2013) noted human factors challenges as trust, control modes (what is the driver allowed to do or not do), and handover of control (how often, when, at what short notice). Schumacher (2013) noted Continental's goal is the adaptation of information, warnings, and vehicle behavior depending on the level of driver distraction. A key aspect under development is a driver analyzer to monitor driver state. This DVI may include a 360° "halo" which directs the driver's attention in the proper direction.

Rupp and King suggested specific DVI techniques (Rupp & King, 2010). Steering characteristics could change as the vehicle approaches a lane boundary, providing feedback on lane position. Subliminal sound could be used to indicate traffic conditions; e.g., as a threat develops, a localized and directional sound would serve to direct the driver's attention. They also discuss the use of augmented reality displays (full windshield or a wearable display) which could highlight objects of interest or other key information, coupled with driver gaze monitoring. They noted:

> Warnings would still have their role as the last resort, but given an immersive situational awareness the driver would be more involved, informed and active in his role, so when it is time to hand over from autonomous to human control it's not a surprise, the context is understood and it will be a mutual decision. (Rupp & King, 2010, p. 14)

Mattern (2013) noted that HMI strategies may require new sensors and functions to facilitate reporting, prioritization, and decision making, interfacing via tones, vehicle dynamics, and personalization to support intuitive system operation, driver acceptance, and trust in the system. Further, he noted the need to clearly indicate what specific driver assistance functions are offered by a vehicle. Interaction with the system can take advantage of smart devices, gesture recognition, capacitive switches, voice recognition, and HUDs to provide augmented reality. New scanning laser technology in particular offers the possibility of larger display areas for HUDs.

Driver Monitoring

Toyota Motor Corporation (2013a, 2013b) described their driver-monitoring system, which constantly monitors the movement of the driver's head when looking from side to side. If the driver's head is turned away from the road when the system detects a probable collision, the system will automatically sound a pre-crash warning alarm. If the situation persists, it will briefly apply the brakes. If the driver does not then respond immediately, all automated emergency braking functions will engage. Toyota Motor Corporation (2012) also reported that the company has funded a 3-year project at Stanford University titled *Driver Vehicle Interface for Partially Intelligent Vehicles*. The research will develop a set of psychological principles that will guide the design of a DVI that provides effective, real-time support for drivers of a partially intelligent vehicle (not further defined). This research will culminate with the development of a DVI for a fully operable automated vehicle to verify its effectiveness.

Ford Motor Company (2012b) described the company's activities to use sensor data from driver- assistance systems, plus biometric data, to develop new methods to estimate driver workload based on traffic and road conditions. The driver workload estimator uses data from active safety sensors plus information on the driver's use of the throttle, brakes, and steering wheel. Side-looking radar sensors used for blind spot monitoring and the forward-looking camera for the LDWS are continuously active, such that the intensity of the traffic situation around the subject vehicle can be estimated. Biometric data sources include several sensors added to the steering wheel rim and spokes to get more detailed driver information, such as the driver's heart rate. Infrared sensors on the steering wheel monitor the palms of a driver's hands as well as his or her face, looking for changes in temperature (a downward- looking infrared sensor under the steering column measures the cabin temperature to provide a baseline for comparing changes in the driver's temperature). Additionally, a sensor is embedded in the seat belt to assess the driver's breathing rate. Ford noted that, with a more complete picture of the driver's health and wellness along with knowledge of what is happening outside the vehicle, the car will have the intelligence to dynamically adjust the alerts provided to the driver and filter interruptions (Ford Motor Company, 2012b). For instance, in heavy traffic, the vehicle control system could increase the warning times for forward collision alerts and automatically filter out phone calls and messages, allowing the driver more time to respond. When the vehicle is on the open road and the driver is assessed as alert, incoming calls could be presented.

In addition, Mattern (2013) noted that awareness of the driver can be facilitated by monitoring eyes, head, pulse, and potentially an "Alive Switch" (i.e., a switch that needs to be triggered by a driver in defined regularity).

DVI Philosophy and Design Relating to Control Transition

Beiker and Calo (2010) discussed research needs with regard to the interaction between drivers and the vehicle when the vehicle navigates autonomously through traffic, especially mixed traffic (i.e., automated and non-automated traffic). Assuming situations will occur such that the automated system requires human intervention, the authors suggested that hand-over scenarios between human and vehicle need to be researched. On a subjective level, they noted that the extent to which humans might feel that their individual freedom increases or is being compromised should be addressed, as well as their perception of trust and safety.

After hand-off to the automated driving function, the driver must maintain vigilance and readiness to resume control, according to Rupp and King (2010). To provide the proper driver support, the DVI must evolve to offer *immersive situational awareness*. Rupp and King cited experiences with automated aircraft cockpits in which operators are uncertain about what the system is currently doing, what it will do next, and what it will do in other similar situations. Three factors are deemed important for an effective DVI (Rupp & King, 2010, p. 9):

1. System will provide timely and specific feedback about the activities and future behavior of the agent relative to the state of the world;
2. User will have a thorough mental model of how the system behaves in particular situations; and
3. System actions are consistent with prior instructions from the human operator.

Per Rupp & King (2010), the automated system could request the driver to confirm readiness to resume control; in the event of a non-responsive driver, it would support the option to bring the vehicle to a suitable non-moving and safely positioned state. What if the driver is deemed unfit to resume control? In this case, the system's task may become that of preventing the operator from starting the car. Therefore, the DVI should include both direct and indirect driver monitoring and interpretation of operator state to ensure properly coordinated driver assist (Rupp & King, 2010).

Rupp and King (2010) concluded their discussion of DVI with a caution. While future DVI designs over the coming decades will guide the transition from *driver* to *operator*, large step changes in DVI design may slow consumer acceptance. Therefore, DVI designs should evolve smoothly and gradually, building upon driver experience with the support and interventions of partial automation.

SECTION 7. LEGAL / LIABILITY ASPECTS TO AUTOMATED DRIVING

This section provides a broad overview of the legal and liability aspects associated with the implementation of automated driving initiatives. The discussion has been divided to address United States perspectives and is followed by a discussion of international perspectives.

U.S. Perspectives

Walker Smith (2013, 2012a) provided the most comprehensive discussion to date of legal issues relating to the sale and use of automated vehicles in the United States. The general conclusion is that the computer direction of a motor vehicle's steering, braking, and accelerating functions without real- time human input is probably legal. However, issues do exist in the regulatory and liability domain. Walker Smith's paper centers on the principle that everything is permitted unless prohibited, and covers three key legal regimes: the 1949 Geneva Convention on Road Traffic, regulations enacted by the NHTSA, and the vehicle codes of all 50 states (2012a). Walker Smith's conclusion is that the Geneva Convention, to which the United States is a party, probably does not prohibit automated driving. The treaty requires every vehicle to have a driver who is "at all times ... able to control" it. He noted this requirement is likely satisfied if a human is able to intervene in the automated vehicle's operation. Walker Smith's analysis further concluded that NHTSA's regulations, which include the Federal Motor Vehicle Safety Standards to which new vehicles must be certified, do not generally prohibit or uniquely burden automated vehicles (2012a).

In tracking state legislative and regulatory actions, Walker Smith (2013) noted that three states (California, Florida, Nevada) and the District of Columbia have enacted legislation related to automated driving. California and Nevada also have considered regulations related to automated driving. In addition, 13 states are beginning to discuss legislative efforts governing automated driving. These efforts are at various stages of the legislative process. Several other states that have not yet implemented legislation regarding autonomous vehicles have brought regulations regarding automation to legislative committee. These states include Hawaii, Massachusetts, Michigan, Minnesota, New Hampshire, New York, Oregon, South Carolina, Texas, Washington, and Wisconsin. Walker Smith also concluded that state vehicle codes probably do not prohibit automated driving, but challenges exist (2012a). These state codes assume that human drivers are using human judgment, and particular rules may functionally require that presence. He observed that many rules mandate "reasonable, prudent, practicable, and safe driving"; these rules have uncertain application to automated vehicles and their users (Walker Smith, 2012a, p. 3). For example, following distance requirements could restrict the operation of close headway vehicle platoons. Walker Smith's paper included draft language for U.S. states that wish to clarify this status (2012a).

Walker Smith recommended five near-term measures that may help increase legal certainty without producing premature regulation (2012a, p. 3):

1. Regulators and standards organizations should develop common vocabularies and definitions that are useful in the legal, technical, and public realms.
2. The United States should closely monitor efforts to amend or interpret the 1969 Vienna Convention, which contains language similar to the Geneva Convention but does not bind the United States.
3. NHTSA should indicate the likely scope and schedule of potential regulatory action.
4. States should analyze how their vehicle codes would or should apply to automated vehicles, including those that have an identifiable human operator and those that do not.
5. Additional research on laws applicable to trucks, buses, taxis, low-speed vehicles, and other specialty vehicles may be useful, in addition to ongoing research into the other legal aspects of vehicle automation.

Soriano (2013) reviewed California Senate Bill 1298 and the California Department of Motor Vehicles (DMV) activities to respond to the legislation. In summary, Senate Bill 1298 calls for the DMV, by January 1, 2015, to adopt regulations for both manufacturers' testing of autonomous vehicles and operation of those vehicles on public roadways. The bill authorizes an autonomous vehicle to be operated on public roads for testing purposes by a driver who possesses the proper class of license and after the manufacturer provides evidence of insurance at the level of $5M; only persons designated by the manufacturer may operate the vehicle. For testing purposes, the driver must be seated in the driver's seat and capable of taking over immediate manual control. In terms of operation by the public, a manufacturer must submit an application to the DMV certifying they meet all testing requirements and safety standards and until DMV approves the application.

Soriano noted that the strategy for development of specific regulations has been split into two packages on differing timescales:

- Regulatory Package 1 (adoption by December 2013)
  - Submission of evidence of insurance
  - Marking of vehicle on DMV's database
  - Other feasible regulations
- Regulatory Package 2 (adoption by December 2014)
  - Testing requirements
  - Safety standards
  - Operator license requirements
  - Vehicle registration requirements
  - Other feasible regulations

In addition to expert working groups and public hearings, a Statewide Steering Committee consisting of the following California agencies provides guidance on a range of issues: DMV; Highway Patrol; Department of Insurance; Department of Transportation; Business, Transportation, and Housing Agency; and the Office of Traffic Safety. The NHTSA also has a seat on the committee.

Walker Smith (2012b) also provided a discussion of the ongoing activity to define terminology relating to automated vehicles, including work by NHTSA, SAE, ISO, and

Germany's Federal Highway Research Institute (BASt). He made the point that a coordinated approach to terminology is an important factor in effective legal, technical, and commercial communication.

Khan, Bacchus, and Erwin (2012) examined levels of automation technology development through the long term (i.e., beyond 2025), noting that increasing automation poses major challenges to government and public policy, as there is no prior experience with the technology. They posit that the advent of automation will require a systematic comprehensive policy framework to address complex issues and also to potentially influence the technology development for societal benefit. This policy framework must be able to deal with uncertainties as well. More specifically, this policy framework would address common technical standards essential for interoperability of systems, safety design standards and regulations, liability and litigation issues, potential incentives for technology adoption, societal benefits, and privacy issues. The authors asserted that the framework should incorporate a methodology to assess automation initiatives to compare costs against the achievement of policy initiatives which include safety, efficiency, mobility, eco-driving, consumer satisfaction, and the needs of special interest groups (e.g., youths, seniors). Since many factors to be included in the automation systems do not have market values or are quantifiable only in subjective terms, a cost-effective method was proposed which captures cost and safety benefits in dollar terms. Benefits that cannot be captured in dollar terms are taken into account by using a utility-theoretic method. Due to uncertainties in cost estimation, costs are expressed in probabilistic terms which take into account empirical data. In their discussion, Khan et al. provided some examples of these key concepts (2012).

Accident Fault

Beiker and Calo (2010) discussed the legal challenges of automated driving. They noted that, while the traditional approach to traffic litigation assumed the cause of an accident to be a human or technical failure, or a result of environmental conditions – or a combination of these – the situation is more complex with automated vehicles. They asserted that the automated vehicle will encounter situations which, in a unique set of circumstances, could contribute to crashes, yet not be found to have had a technical failure per se. It is unclear how the courts, or the public, will respond to the prospect of artificial intelligence acting on behalf of humans with life or death consequences. They see a danger of overreaction, even if it can be established that a transition to automated vehicles would lead to far fewer traffic-related fatalities overall. Beiker and Calo (2010) contended that mitigating this issue will require research and education, including methods such as:

- Pilot fleet communities with statistical comparisons
- Extensive beta testing with "limited autonomy"
- Mock trials and focus groups
- Special insurance policies for automated vehicles
- Mandatory data recorders for automated vehicles

They expect that a set of policies can be established to create the necessary legal framework for further development of vehicle automation.

Kalra, Anderson, and Wachs (2009) evaluated how the existing liability regime (i.e., traditional negligence, no-fault liability, and strict liability) would likely:

- Assign responsibility in crashes relating to vehicle automation,
- Identify the controlling legal principles at work, and
- Examine implications for further development and acceptance of the technology.

They found that the existing liability situation does not appear to create unusual liability concerns for owners or drivers of vehicles equipped with automation. Manufacturers' product liability, however, is expected to increase, as they may be held responsible under several theories of liability for systems that aid the driver but leave him or her in total or partial control.

In particular, current liability law on design defects may hinder the adoption of automated vehicle technologies. Kalra, Anderson, and Wachs (2009) provided the example of an automatic emergency braking system which works to prevent crashes 80 percent of the time; however, the other 20 percent of the time, the technology does not work, and crashes occur as if the technology were absent. Victims in those crashes may sue the manufacturer and argue that the product was defective because it failed to operate properly in their crashes. This argument could be valid under the existing liability doctrine, if it could be established that the product did not work as designed (manufacturing defect), as advertised (misrepresentation), and as warranted. Even though the social benefits are strong, a manufacturer may delay bringing this type of technology to market. Kalra, Anderson, and Wachs asserted that the existing liability regime does a poor job of aligning private incentives with the public good in a situation such as this (2009).

In the event of manufacturer reluctance to bring advanced driver-assistance systems to market, Kalra, Anderson, and Wachs cited consumer education as an important factor in the proper use of the systems (2009). They also posited that federal regulations could preempt state tort suits if the USDOT establishes regulations pertaining to vehicle automation. Another approach offered is to more fully integrate a cost-benefit analysis into the standard for liability in a way that accounts for the consideration of the benefits associated with this technology; here, more research is needed to fully capture costs and benefits.

Kalra, Anderson, and Wachs additionally reviewed the existing literature on the regulatory environment for automated vehicle technologies (2009). They found that, to date, no government regulations exist for these technologies. For industry-led standards to work, they stressed the importance of precisely specifying environmental conditions under which compliance must be met, as well as addressing how diverse populations of drivers may use the technologies. Because drivers are likely to use these technologies in different vehicles created by different manufacturers, they called for standardization of system performance and user interfaces. They asserted that safe use depends on the driver understanding how to use the system as well as understanding its capabilities and limitations (Kalra, Anderson, & Wachs, 2009).

Further, Rupp and King (2010) hold that the driver has legal responsibility for control of the vehicle and must have the ability to override the system via steering, brake, or throttle. This would include the option to request or make maneuvers such as a lane change. The driver may also be requested to confirm appropriateness and acceptance of a system-recommended maneuver.

European Perspectives

VDA (i.e., the German automakers association) established a working group on automated driving (Bartels, 2013). Major OEMs and Tier One suppliers are participating. The objective is to create a framework of conditions for establishing the automated driving function. Their focus is on coordination of activities regarding definitions and terminology, international homologation (i.e., the United National Economic Commission for Europe [ECE] Vehicle Regulations), and regulatory law (e.g., The Vienna Convention on Road Traffic 1968, German Road Traffic Code).

Van Dijke and van Schijndel (2012) provide an overall review of the European CityMobil project, which ended in 2011. They noted that the most severe barriers to implementation were legal and administrative. In the legal domain, current European legislation (based on the Treaty of Rome) requires that the driver be responsible at all times for a vehicle that uses public roads. This requirement raises numerous questions for automated vehicles where there is no driver, such as responsibility in a crash. The authors made the point that the uncertainty is viewed as effectively barring automated systems from using public roads, thus restricting them to private terrains (van Dijke & van Schijndel, 2012). Changes to the existing law are expected to take a significant amount of time.

In the safety and certification domain, van Dijke and van Schijndel (2012) maintained that the key issue is not so much doubt about the safety of automated systems, but the lack of established criteria and methods to certify a system as *safe enough*. Although such procedures exist for traditional roadway vehicles, such standards do not yet exist for automated transport systems of the type developed in the CityMobil project. To get permission from local authorities for system demonstrations, certification procedures were developed, mainly relying on a Failure Modes, Effects, and Criticality Analysis (i.e., FMECA) process. For a CityMobil demonstration in Rome, the Italian Ministry of Transport accepted the results as the main basis to certify the system. For another demonstration in Abu Dhabi, the Ministry of Transport required an additional analysis before the system was accepted. The authors noted that the process of obtaining certification was time-consuming and highlighted the importance of establishing generally accepted procedures. A next step envisioned was to introduce a set of procedures to the European authorities as a possible future certification standard for automated transport systems (van Dijke & van Schijndel, 2012).

SECTION 8. TIMELINE OF VEHICLE AUTOMATION

Not surprisingly, a variety of viewpoints exist as to when specific levels of vehicle automation will become available. Some see very rapid rollout. Silicon Beat (2012) quotes Google cofounder Sergey Brin as saying, "You can count on one hand the number of years until people can experience this." Keane (2013) quotes Anthony Levandowski, product manager for Google's self-driving car technology, as saying, "We expect to release the technology in the next five years. In what form it gets released is still to be determined."

Auto industry players generally align in their expectations. AUVSI (2012) reported on comments from industry executives made at the Driverless Car Summit 2012. Dr. Gary Smyth of GM expects "transferred control" (hands/feet off driving) (L2) by the mid-end of the decade and automated driving by the end of the decade. Similarly, Bigman (2013) quoted Nissan CEO Carlos Ghosn at the Detroit Auto Show stating that automated driving will be ready "by the end of this decade." In addition, Luca Delgrossi of Mercedes spoke to Daimler's stepwise approach to automated driving. Daimler is looking at automated driving as their final goal and is going step by step to get there. The Daimler progression calls for moving from *feet off* (today's ACC) to add *hands off*, which would require short driver takeover times. The next step would be eyes off requiring moderate takeover times. The ultimate goal would be *body out*, meaning an unoccupied vehicle. Delgrossi noted that *hands off*, *eyes off*, and *body out* are not certifiable today.

At the more conservative end of the scale, Beene (January 2013) reported on comments made by industry executives at the Automotive News World Congress. Helmut Matschi of Continental AG said automated driving systems will be able to take over for drivers by 2025. He expects TJA will be possible by 2016 and that hands-free driving, which allows drivers to have some time to resume manual control of a vehicle when alerted will be possible by 2020. Ludwig Willisch, CEO of BMW North America, said automated driving technologies will likely enter the market before 2025.

Ford's Blueprint for Mobility (Bunkley, 2012) provided their vision for the rollout of automated driving other the next several decades.

- Near term (2012–17)
  - Implement limited automated functions for parking and driving in slow-moving traffic, including active park assist, ACC, and Active City Stop, which automatically applies the brakes if a crash is about to occur.
- Midterm (2017–25)
  - Introduction of semi-automated driving technologies, including driver-initiated "auto pilot" capabilities and vehicle platooning in limited situations.
  - Increasing capabilities of driver-assist technologies, including limited semi-automated and automated highway lane-changing and exiting.
- Long term (2025–30)
  - Arrival of fully automated assist capability, plus the arrival of automated valet functions, in which vehicles park and retrieve themselves.
  - Letting the car take over.

Oonk and Svensson (2013) reported that the draft roadmap for the European iMobility Forum Working Group on Vehicle Automation places "urban platooning" in the 2019–2020 timeframe and "full automation" in the 2025–2030 timeframe.

While not providing specific timeframes, Rupp and King (2010) offered a useful sequence of increasing system capability. Specific cases are listed here, along with variations.

- Steady State Control: Limited automated control for a short interval in a specific driving scenario. This control level could be a function like ACC in a single axis. A further extension of this use case would be limited automated control across multiple axes: such as TJA or driving on the highway from entrance ramp to exit ramp, possibly including lane changes. This functionality might be limited to roadways that the vehicle has previously driven and analyzed to be "self-drivable" in terms of GPS availability, number of lanes, etc. The system may still ask the driver for confirmation for certain maneuvers, such as a lane change. Automatic parking and platooning would fit into this use case.

- Transitional Control: This use case addresses scenarios in which vehicles may conflict in intent and space, with support provided through information, advice, warning, or automatic control, ideally as early smooth coordination or as late evasive actions, if necessary. Sub-categories are:
  - Merging at freeway and intersection flow points
  - Assistance at intersections with opposing flow traffic
  - Convenience support at an intersection, such as automatic slowing and stopping for a stop sign or traffic signal
  - Emergency stop on the roadside in case the driver is incapacitated; this would include communicating the emergency situation to surrounding traffic
- Revisiting Known Destinations and Routes: This use case extends to all roads but is still restricted to roadways that the vehicle has already traversed and passively assessed. In this case, the vehicle only has to confirm interactions with the environment, rather than analyze the entire situation in real time. Subcategories are:
  - Areas frequently traveled, for example, from home to work, such that the system has high confidence in familiarity and low likelihood of change in the nature and condition of the infrastructure and the typical traffic flow.
  - A vacation or holiday destination, which is a longer distance and less frequently traveled, introducing a greater situational variability since the last traversal and increasing demand on the automated system to recognize changes in the infrastructure and traffic flow characteristics.
  - A fully automated local shuttle scenario described as "a limited pre-implementation feasibility demonstration and learning opportunity only." The shuttle would function as a test bed where the new DVI, situational awareness, and control functionality can be assessed for reliability and robustness. They posit that this type of shuttle could operate on a private road network.
- Traversing Unknown Routes: The General Case represents full automated vehicle functionality, extending to situations not previously assessed. The vehicle is capable of handling all scenarios. Rupp and King noted (2013) that full functionality should achieve "at least the same outcome as the human driver when encountering new situations, but with the greater diligence and situational awareness, as well as rapid recognition of subtle novelty that a machine can have." This full capability would include any configuration of intersection, operating in the full range of weather conditions, operating on poorly marked or non-paved roads, and handling scenarios which require evasive action in traffic, even including the possibility of assessing and using driving surfaces off- road.

Rupp and King concluded by noting that the evolution to this end state will parallel the evolution from the horse-drawn to the automobile carriage, which they characterize as:

> a period of initial caution and low acceptance, initial innovation and invention, use by early adopters, followed finally by rapid innovation and expansion, mass market penetration, and standardization.

Full consumer acceptance will depend on consumers observing early adopters long enough to build trust in automated systems (Rupp & King, 2010).

Schumacher (2013) offers the following timeline, which follows the definitions of Levels 2, 3, and 4, respectively:

- 2016 – Partially Automated: Monitoring of the system required; the driver needs to be able to take over the driving task at any moment. Example: S & G ACC up to approximately 56 km/h (30 mi/h)
- 2020 – Highly Automated: Monitoring of the system is not required; the driver needs to be able to take over the driving task with lead time. Example: S & G ACC (highway)
- 2025 – Fully Automated: Monitoring of the system is not required; the driver does not need to take over the driving task. Example: Highway driving up to 130 km/h (approximately 81 mi/h)

SECTION 9. SUMMARY

This document seeks to provide a summary of past research work, a view of the current state of automation technology, and information on emerging system concepts. This summary is provided within the context of L2 and L3 automation concepts. It is, however, important to note that the literature review performed as part of this task is not meant to be a comprehensive effort; instead it expands upon previous and recent literature reviews performed on this topic for the USDOT.

Government-sponsored research programs have been a major force behind the development of some L2 and L3 automated driving concepts. European and Asian government research efforts have furthered development of concepts such as vehicle platoons, speed adaptation, automation for personal mobility, and general L2/L3 vehicle operations. In addition, a number of automakers and suppliers are either preparing to or actively working towards fielding L2 and L3 automated vehicles in the near future. The motivation for automakers is centered on issues of safety, convenience, enabling multitasking, and increasing mobility. These motivations are also reflected in the ongoing development of systems such as TJA and various highway-speed automated driving aids.

In addition to studies directly addressing automated driving, this effort has sought out studies not directly focused on automation yet dealing with tasks and scenarios relevant to automated driving. Further, information regarding differing policy approaches has been summarized in order to provide a more relevant context for the state of vehicle automation. Automated driving-relevant databases are available; this document identifies databases of this type in support of future research efforts. Finally, without attempting to predict the future actions of the market, regulatory bodies, or manufacturers, the timeline of vehicle automation was provided in the interest of summarizing discussions and publicly released information on forthcoming L2 and L3 automated vehicles.

As the primary focus of this review was to perform an initial Human Factors assessment of driver performance and behavior, the remaining portion of this section will draw attention to the key findings associated from the literature. Additionally, this discussion includes an overview of the findings resulting from the review of articles, summaries, and presentations from OEMs and suppliers which shed additional light on their roadmap for automation and philosophy for driver role.

Studies of Automated Driving

Government-sponsored research programs play an important role in stimulating research and development and policy initiatives in vehicle automation. Key programs and findings related to the current effort were reviewed. More in-depth discussions about many of the projects presented can be found in the recently released literature reviews of international activity in cooperative vehicle-highway automation systems (Shladover, 2012a, 2012b).

Studies of automated driving have demonstrated that automated vehicles have the potential to aid in overall driving performance. In the military setting, automated vehicles have proved successful in convoy situations (Schoenherr, 2009) and have been shown to improve gap distance maintenance and improved panic stopping distances over non-automated systems (Thiessen, 2011). Additionally, researchers found that automated systems resulted in significantly less participant fatigue and improved performance (Theisen, 2011). When used for patrol operations, automated vehicles have replaced soldiers on the front lines, possibly reducing soldiers' exposure to gunfire or other confrontations (Main, 2013). Safety benefits are also seen in private industry, where automated vehicles have been used in mines as haul trucks (Rio Tinto, 2012). In the academic setting, several efforts, including the University of Braunschweig in Germany's Leonie vehicle, the VisLab Intercontinental Autonomous Challenge, the Oxford University RobotCar project, and work completed at the Karlsruhe Institute of Technology have focused on developing an automated vehicle that incorporates highly integrated sensors and cameras in order to provide a low-cost solution (Northdurft, 2011; Broggi et al., 2012; Bertozzi et al., 2011; Lee, 2013; Karlsruhe, 2013; Lenz, 2013). In a simulator setting, Neubauer et al. (2012) assessed the impact of L3 automation on fatigue, stress, and workload and suggested that additional solutions which maintain driver engagement and address the vulnerabilities of fatigue-prone drivers should be sought.

Research Approaches Supporting Automated Driving

More extensive work has been performed examining what may be considered underlying Human Factors technologies that could be considered as part of an L2 or L3 automated system. These efforts include the DVI design and evaluative approaches of IVIS, especially with regard to such systems' effect on driver workload. Further efforts are associated with the development of ACC and FCW systems, lane maintenance and LDWS, and connected vehicle initiatives. The following key Human Factors issues related to these enterprises were identified.

IVIS and Associated DVI Issues

While distraction is a concern, when designed properly an IVIS can aid driving. Davidse et al. (2009) found that driver performance in both older and younger drivers improved when provided with information on right-of-way regulations, obstructed intersection views, and safe gaps to cross or join traffic streams, resulting in safer driving performance without increasing workload. A key aspect is maintaining situational awareness and monitoring situational developments when interacting with in- vehicle devices (Rauch et al., 2010; Reimer et al., 2010, Merat et al., 2012). Kaber et al. (2012) explored the effects of visual, cognitive, and simultaneous visual and cognitive distraction on operational (braking, accelerating) and tactile (maneuvering) tasks and found that the three types of distraction increased workload. They found that drivers were better able to adapt to visual distraction through increased headway while tactical control behavior was less conducive to adaptation. However, drivers engaged in tactile control were able to perceive the higher workload when presented with simultaneous distractions, and, as a result, prioritize the driving task. When looking at the effects a secondary task has on primary-driving-task performance degradation, several have found it beneficial to use the lane change task (Benedetto et al., 2011) or a lane change task adapted to automated driving conditions (Spießl, 2011; Spießl & Hussmann, 2011).

A number of research efforts have focused on information presentation. Two separate meta-analyses efforts explored the use of tactile and auditory interruptions (Lu et al., 2013; Lu et al., 2011). Researchers found that tactile interruptions were responded to 6% faster than auditory interruptions and resulted in no significant differences in task performance and no speed-accuracy tradeoff. When taking into account moderator variables, they recommend using tactile cues for low-complexity IT conditions and for notification alerts and using auditory cues for high-complexity IT conditions and urgent alerts. Much research has focused on the use of force-feedback (haptic) shared control as a DVI that can provide for shared control between human-automation interactions (Abbink et al. 2012). Research has shown that haptic shared control can lead to short-term performance benefits, including improved car following behavior (Abbink et al., 2008), curve negotiation (Mulder et al., 2012; Mulder et al., 2008; Mulder & Abbink, 2010), lane keeping (Tsoi et al., 2010), collision avoidance maneuvers (Penna et al., 2010; Brandt et al., 2007), and task performance (Griffiths & Gillespie, 2005).

Trust in Automated Vehicles

Several studies have examined drivers' trust and acceptance of automated systems. Systems that were able to both take control over vehicle actions and also provide information to the driver about the driving goals were judged to be more trustworthy and acceptable than systems that take control without providing information (Verbene et al., 2012). Similarly, communicating situations in which the automated system is uncertain about a situation was found to improve users' trust in automation and system acceptance (Beller et al., 2013). As compared to the impact that trust and satisfaction that aesthetics plays in mobile commerce or websites, initial research indicates that system aesthetics has little impact on users' trust and acceptance of automated systems (Weinstock et al., 2012).

Researchers have also explored trust associated with automated system accuracy (Weinstock et al., 2012) and reliability. Findings suggest that initial perceptions of reliability levels impact both subsequent reliability estimates and trust ratings (Blair et al., 2012). Results indicated that false alarms diminish trust and compliance, whereas the context associated with unnecessary alarms fostered trust as compliance during subsequent events (Lees & Lee, 2007). Additionally, findings suggest that providing drivers with continuous information about the state of automation may be more effective than warning drivers of imminent crash risk when the system fails (Seppelt & Lee, 2007). Increasing driver education as to system functionality and limitations may be warranted (Larsson, 2010). As drivers become more aware of system limitations, they may actually interact with the system more, thus keeping them in the loop (Larsson, 2012). Following gaps also affect drivers' opinions of the automated systems. For example, the PATH research found that participants' were generally receptive to CACC following distances that were shorter than 1s. When compared to manual driving preferences, the time-gap settings offered by ACC systems resembled time gaps participants would keep while driving manually in light to medium traffic, while CACC systems were closer to the time gaps that drivers would keep while driving manually in heavy traffic (Shladover et al., 2009; Nowakowski et al., 2010; Shladover et al., 2011).

ACC, FCW, and Connected Vehicles

Key studies explored underlying driver behavior and performance issues that arise from such advanced technologies as ACC, FCW, and LDWS. Several large-scale FOTs have been conducted in this area. The RDCW FOT observed a 10- to 60-percent reduction in departure conflict frequency at speeds over 88 km/h (55 mi/h) when the system was activated. Participants also gave the following features favorable ratings: LDW and CWS alert times, LDW and CWS missed alert frequency, and LDW false-positive alert frequency (Wilson et al., 2007). Results of the ACAS FOT indicated a reduction in the incidence of tailgating as compared to unassisted drivers. In terms of driver acceptance, the ACC system acceptance was uniformly high, but mixed for the FCW, possibly owing to the reduction in workload and stress afforded by the ACC and the credibility issues resulting from the false alarms generated by the FCW system (GM, 2005). The CAMP FCW provided additional insight into the development of a DVI to accompany an FCW, how driver behavior and judgments compared under on-road versus simulated driving approach conditions, and recommendations for simulated scenario design (Keifer et al., 2005). Findings included the identification of a single-stage, dual modality (auditory and visual) alert that was found to be robust, effective, and judged appropriate across a wide range of tested conditions.

In addition to the large-scale FOTs, further efforts inform the knowledge base regarding the effects of stress and mood, reaction times in critical situations, use patterns of early adopters, driver following, and headway patterns. Funke et al. (2007) explored the effects of stress, vehicle automation, and subjective state on driver performance and mood. The findings suggested the potential benefits that the DSI screening tool may have for driver personnel selection efforts and the need to educate drivers, especially inexperienced drivers, about the negative consequences of stress. Vollrath et al. (2011) found that although drivers did not shift more attention to secondary tasks when driving with ACC, drivers did exhibit delayed reactions in critical situations. Among early adopters of ACC, safety implications were associated with individuals' level of understanding, level of trust in automation, and driving style and personality. An evaluation of in-car displays (i.e., an iconic display, a flashing iconic display, and a representation of the radar data) used to support S&G ACC found that drivers correctly identified more changes detected by the system with the radar data display than with the other displays; however, the higher detection of workload accompanied the increased detection (Stanton et al., 2011). Naturalistic driving results suggested that the presence of a warning in an IVBSS resulted in increased headways and faster responses to forward conflicts. Saffarian et al. (2012) also looked at driving headways, in this case those associated with fog conditions in automated and manual driving scenarios, with findings indicating that the two main advantages in maintaining close headway in fog are reduced perceived risk and improved lateral control. Further, no differences in risk perception were found between manual and automated driving when following distances were taken. When looking at the effects of time-gap settings and contents of secondary tasks on bus drivers' performance when regaining control from ACC in a car-following scenario during expressway driving, Lin et al.'s (2009) results indicated that safer time gaps were longer than 1.60 s for non-secondary task distraction and longer than 2.08 s for being continuously distracted by secondary tasks. Situational awareness benefits were associated with the alerts presented in the SAFE-Trip 21Initiative, with drivers finding the alerts most useful when traffic cues appeared unexpectedly.

While these findings are informative, it should be noted the majority of the studies included a limited number of participants and involved simulated or test conditions. Even in the case of real-world applications, longer-term studies would provide further insight as to the effects of automation over time. As such, many of the researchers included above noted the need for future research (e.g., Thiessen, 2011; Davidse et al., 2009; Abbink et al., 2012; Weinstock et al., 2012; Lees & Lee, 2007). An associated issue to be explored relates the education provided with the system, whether additional training is necessary and what effects additional information and/or system experience may have on system understanding and use (e.g., Larsson, 2010; Funke et al., 2007; Vollrath et al., 2011).

Lessons Learned from Other Domains

Several studies provide insights for systems engineers, particularly in regard to adaptive and co- adaptive systems. Feigh et al. (2012) provided a systemic framework that categorized the ways in which adaptive systems can modify their behaviors as well as trigger mechanisms through which adaptive systems can sense the current situation and decide to adapt. Christensen and Estepp (2013) expanded upon the theory of adaptive aiding by measuring the effectiveness of co-adaptive aiding, concluding that a third day of testing extended the time period which provided enough time and experience for user adaptation as well as online system adaptation. Niederée et al. (2012) concluded a need exists to focus on automation's observability when designing highly automated systems, especially in safety-critical domains. Cognitive countermeasures were found to be effective at mitigating excessive focus issues in the UGV environment (Dehais et al., 2011). Further, the principle of cognitive countermeasures could be applied to a large domain of applications that involve human operations interaction with critical systems. The use of memory stores for controlling an adaptive system was found to provide insight into the impact of cueing of mode transitions which can also inform future system designers (Kaber & Kim, 2011). Rovira and Parasuraman (2010) found that conflicts were detected faster and more accurately with reliable automation than with manual performance; however, even highly reliable yet imperfect automation resulted in serious negative effects on operator performance. Therefore, system designers were urged to provide users with feedback on the state of the automation or other tools that allow for inspection and analysis of the data underlying the conflict probe algorithm.

Imperfect automation systems were also found to be detrimental within the context of a simulated military multitasking environment. In this case, false-alarm-prone automation errors were more detrimental to those with higher perceived attentional control while miss-prone automation errors were more harmful for those with lower perceived attention control (Chen & Barnes, 2012). Further, visual cueing was preferred by those with low spatial ability while those with high spatial ability favored tactile cueing. These findings were further supported by Chen and Terrance (2009), who found similar cueing preferences. Within the management of a team of ground robots, researchers found that participants' attentional control and video game experience affected their overall multitasking performance and that participants with greater spatial ability consistently outperformed low-spatial-ability counterparts in tasks requiring effective visual scanning (Chen & Barns, 2009). In terms of display, Neyedli et al. (2011) evaluated display formats for automation combat identification aids and found that integrated display and mesh display formats were the most effective.

Military-based research also provided insight into the effects of automation on task performance. Distraction due to boring, low-task-environments were found to be manageable through efficient attention switching; however, additional research is needed to determine the frequency and duration of attention state switches (Cummings et al., 2013). For those human operators supervising multiple UAVs and UGVs under high-workload conditions, adaptive automation (i.e., automation in which individual operator change detection performance was assessed in real time and used to invoke the ATR only when change detection was below a threshold) was shown to have beneficial effects (Parasuraman et al., 2009). Findings also suggested that overload of information may be an issue for UAV interface designers (Guznov et al., 2011).

In the rail domain, research has focused on the impact of automation on human resource management. Cunningham (2007) noted that within the rail industry efforts implemented to aid in the management of human performance include psychometric testing, limited workdays, regular breaks, and fatigue education. Vlad and Tatarnikov's (2011) findings also support the need for fatigue education. In a small qualitative study, Balfe et al. (2012) found that staff developed coping mechanisms to compensate for generally low levels of understanding about the automated system and an inability to predict automation system actions. Additionally, Pickup et al. (2010) developed a tool, the ODEC, for determining mental workload in signal operators for human resource management purposes; this tool may have applications in other domains.

Control system automation research provided insight on the effects of individual differences which may have implications for user acceptance and system reliance. McBride et al. (2010) found that in a dual-task scenario, younger adults outperformed older adults and exhibited less dependence on the automation and took significantly less time to verify automation suggestions than older adults who reported greater trust in the automation and experienced higher workloads. Jipp (2012) explored individual differences in fine motor skills and found performance to be negatively affected by those with lower fine motor skills. The effects of the influence of implicit attitudes towards automation were also explored (Merritt et al., 2012). Researchers concluded that implicit attitudes have important implications for trust; users may not be able to accurately report why they experience a given level of trust; measurements of explicit and implicit predictors may be necessary to understand why users trust, or fail to trust, automation. Mood also influences reliance on and trust in automated systems (Merritt, 2011). Happiness significantly increased trust and liking for the automated system. Furthermore, liking a new system may be key to appropriate automation reliance. As Merritt notes, positive affect can be easily induced; therefore it may be a lever for increasing liking. Reichenback et al. (2011) noted the benefits of automation as countermeasures. Automation was found to help protect performance after a period of wakefulness. Also, those suffering from sleep loss reallocated resources and showed more attentive behaviors towards possible automation failures.

Industry Activity

This literature search found that the majority of automakers active in the United States, one truck OEM, and Google have made public statements as to their ongoing development of automated driving systems. Some stress safety as foremost, others also note the ability for drivers to do other things while driving as a key benefit of this technology. Several note the value of automated driving to extend mobility for the disabled and elderly.

The majority of product-oriented industry activity focuses on L2 automation in highway driving. In the low-speed domain, model year 2014 Traffic Jam Assist products from two OEMs have been announced, which include limited abilities at regular highway speeds as well. Several OEMs and suppliers are actively testing full-capability L2 systems (including automatic lane changing) on public roads at highway speeds to evaluate the robustness of the systems. For L3 automation, the need for redundancy in signals, software, and electrical systems has been noted since the driver role is diminished. Activity at Level 4 automation focuses on implementation of an emergency stop assistant as well as a "commuting system." Public road testing is underway for all these levels.

For automation in urban driving environments, the majority of published work examining city driving has come from Europe. In many cases, this builds on prototypes created for the DARPA Urban Challenge. Systems under development and test aim for full situational awareness and proper behavior on multi-lane city streets, narrow streets, and at intersections, detecting and responding appropriately to other traffic and pedestrians. Special-purpose urban vehicles will be tested in European cities in the near term, and one commercial urban vehicle for pedestrian zones has entered the market.

Automated valet parking is a topic of high interest as well and could serve as a near-term stepping stone to on-road automation.

Technology Challenges

A majority of the activity in developing technology to accomplish automated driving focuses on situational awareness and decision-making, as well as on overall system robustness and test methods. Key challenges and requirements for automated driving were noted as:

- Providing fault-tolerant/fail-safe automated vehicle control (with driver-in-the-loop)
  - Ability to handle faults so that systems are "fail-operational"
- Maintaining situational analysis in complex environments
  - Object detection capability to minimize false negatives
  - Detection of all relevant road elements
- Addressing emergency situations and rare events
- Integration of V2X communications including security/privacy, interoperability, signal congestion
- Low sensitivity to weather and lighting conditions
- Physical redundancy

Some view V2V and V2I network communications as an additional sensor that will improve the accuracy and timeliness of information when fused with other onboard sensors, extending sensing beyond line of sight and providing information about traffic, weather, road condition, and states of nearby vehicles.

Several sources noted the need for robust safety testing, including on-road testing in the range of millions of kilometers. Experts have observed that specific development and test procedures over and above current methods used for active safety systems are needed for testing automated driving.

User Issues

Does the public want automated driving? Consumer surveys have reported numbers as high as 49% and as low as 18% favorable towards having an automated vehicle. Males, owners of premium vehicles, and those living in urban areas were found to have higher interest. Based on customer clinics in the European SARTRE project, project engineers concluded that automated driving is easy to adapt to "for most people" (Coelingh, 2013). Further, they concluded that half of today's car buyers are ready to embrace autonomous driving and that "a majority of tomorrow's car owners will not buy a car without it" (Coelingh, 2013).

Potential user issues with automated vehicles include, but are not limited to, the following. The importance of the driver having a clear understanding of system modes was noted. The ability to use driver state sensing to monitor the focus of the driver's attention plus the driver's intention is of high interest and development is underway by several OEMs. As one view, an active partnership between the driver and the vehicle system was described, key factors are coordination (sharing information), cooperation (aware of and support for each other's goals), and collaboration (working on a shared project). Work is also underway to develop "cognitive automobiles" which integrate probabilistic reasoning and knowledge of human driving to make behavior decisions. For complex traffic scenarios, one aim of the current research is to intelligently filter information to avoid overloading the driver, as well as to adapt information presentation to the driver's current state; ideally this will result in the driver receiving information much earlier, thus allowing him or her to anticipate traffic situations rather than simply react to them.

Legal / Liability Aspects

United States

Legal experts have generally concluded that the computer direction of a motor vehicle's steering, braking, and accelerating functions without real-time human input is probably legal (Walker Smith 2013; 2012a). Further, the Geneva Convention, to which the United States is a party, probably does not prohibit automated driving. Additionally, NHTSA's Federal Motor Vehicle Safety Standards do not generally prohibit or uniquely burden automated vehicles.

Three states (California, Florida, Nevada) and the District of Columbia, have enacted legislation related to automated driving. California and Nevada also have considered regulations related to automated driving.

Walker Smith recommended five near-term measures that may help increase legal certainty without producing premature regulation (2012a, p. 3):

- Regulators and standards organizations should develop common vocabularies and definitions that are useful in the legal, technical, and public realms.
- The United States should closely monitor efforts to amend or interpret the 1969 Vienna Convention, which contains language similar to the Geneva Convention but does not bind the United States.
- NHTSA should indicate the likely scope and schedule of potential regulatory action.
- States should analyze how their vehicle codes would or should apply to automated vehicles, including those that have an identifiable human operator and those that do not.
- Additional research on laws applicable to trucks, buses, taxis, low-speed vehicles, and other specialty vehicles may be useful, in addition to ongoing research into the other legal aspects of vehicle automation.

A great deal of attention is now focused on the California regulatory process for automated driving. Soriano (2013) reviewed California Senate Bill 1298, which calls for the California DMV to adopt regulations for both manufacturers' testing of autonomous vehicles and operation of those vehicles on public roadways by the beginning of 2015. This work is proceeding in two phases to meet the legislature's deadline.

Khan, Bacchus, and Erwin (2012) noted that increasing automation poses major challenges to government and public policy, as there is no prior experience with the technology. They expect that the advent of automation will require a systematic comprehensive policy framework to address complex issues such as technical standards, potential incentives for technology adoption, and societal benefits.

As to liability, Beiker and Calo (2010) noted that the situation is more complex with automated vehicles, concluding that it is unclear how the courts, or the public, will respond to the prospect of artificial intelligence acting on behalf of humans with life or death consequences. They expect that a set of policies can be established to create the necessary legal framework for further development of vehicle automation.

Kalra and Wachs (2009) found that the existing liability situation does not appear to create unusual liability concerns for owners or drivers of vehicles equipped with automation. Manufacturers' product liability, however, is expected to increase, as they may be held responsible under several theories of liability for systems that aid the driver but leave him or her in total or partial control. They also posited that federal regulations could preempt state tort suits if the USDOT establishes regulations pertaining to vehicle automation. Another approach offered is to more fully integrate a cost-benefit analysis into the standard for liability in a way that accounts for the consideration of the benefits associated with this technology; here, more research is needed to more fully capture costs and benefits.

Europe

VDA (i.e., the German automakers association) established a working group on automated driving (Bartels, 2013). Major OEMs and Tier One suppliers are participating. The objective is to create a framework of conditions for the establishment of the automated driving function. Their focus is on coordination of activities regarding definitions and terminology, international homologation (i.e., the United National Economic Commission for Europe [ECE] Vehicle Regulations), and regulatory law (e.g., The Vienna Convention on Road Traffic 1968, German Road Traffic Code).

Van Dijke and van Schijndel (2012) make the point that the uncertainty is viewed as effectively barring automated systems from using public roads, thus restricting them to private terrains. Changes to the existing law are expected to take a significant amount of time. With regard to CityMobil urban vehicles, they noted that a next step envisioned was to introduce a set of procedures to the European authorities as a possible future certification standard for automated transport systems (van Dijke & van Schijndel, 2012).

Timeline

Auto industry players generally align in their expectations as to when specific levels of vehicle automation will become available. Assimilating various comments from industry players, the following timeline is representative:

- 2016 (Level 2): combined lateral and longitudinal control with monitoring of the system required; the driver needs to be able to take over the driving task at any moment. Stop-and-go driving on a limited access highway at low speeds is frequently mentioned.
- 2020 (Level 3): highly automated driving at highway speeds in which active monitoring of the system is not required; the driver needs to be able to take over the driving task with some reasonable lead time.

- 2025 (Level 4): fully automated driving on highways in which monitoring of the system is not required and the driver does not need to take over the driving task.

While researchers are actively developing automated systems for non-highway urban street environments, at this time no concrete statements have been made as to when this capability might be commercially available.

APPENDIX A. LIST OF ACRONYMS

ACC	Adaptive Cruise Control
ACAS	Automotive Collision Avoidance System
ADAS	Advanced Driver Assistance Systems
ADPTS	(The Netherlands) Advanced Public Transport Systems
ADS-B	Automatic Dependent Surveillance - Broadcast
AHS	Advanced cruise-assist Highway System (AHS)
AiTR	Aided Target Recognition
ALCT	Autonomous Lane Change Test
AMAS	Autonomous Mobility Appliqué System
AQuA	Automated Queue Assistance
ARC	Automated Assistance in Roadwork and Congestion
ARS	Automatic Route Setting
ASAS	Airborne Separation Assurance System
ATC	Air Traffic Control
ATM	Air Traffic Management
ATR	Automatic Target Recognition
ATSP	Air Traffic Service Providers
BASt	Bundesanstalt für Straßenwesen
CACC	Cooperative Adaptive Cruise Control
CAMP	Crash Avoidance Metrics Partnership
CAN	Controller Area Network
CAST	Convoy Active Safety Technology
CC	Cruise Control
CDTI	Cockpit Display of Traffic Information
CIB	Collision-imminent Braking
CMB	Collision Mitigating Braking
CR&D	Conflict Resolution and Detection
CSW	Curve-Speed-Warning Subsystem
DALI	Driver Activity Load Index
DARPA	Defense Advanced Research Projects Agency
DMV	Department of Motor Vehicles
DGPS	Differential Global Positioning System

DSA	Driver State Assessment
DSI	Driver Stress Inventory
DSRC	Dedicated Short Range Communications
DSSQ	Dundee Stress State Questionnaire
DVI	Driver-Vehicle Interface
EC	European Commission
EEG	Electro Encephalography
EID	Ecological Interface Design
EKG	Electrocardiogram
EOG	Electrooculography
ERASMUS	En Route ATM (Air Traffic Management) Soft Management Ultimate System
ESA	Emergency Stopping Assistant
FAA	Federal Aviation Administration
FCW	Forward Collision Warning System
FCW+	Forward Collision Warning and Braking System
FLR	Forward Looking Radar
FMECA	Failure Modes, Effects, and Criticality Analysis
FMS	Flight Management Systems
FOT	Field Operational Test
GM	General Motors Corporation
GOMS	Goal, Operators, Methods, and Selection model
GPS	Global Positioning System
HAD	Highly Automated Driving on Freeways
HAVEit	Highly Automated Vehicles for Intelligent Transport
HMI	Human-Machine Interface
HUD	Heads-Up Display
I2V	Infrastructure To Vehicle
ICA	Integrated Cruise Assist
ICT	Information and Communication Technologies
ISi-PADAS	Integrated Human Modelling and Simulation to support Human Error Risk Analysis of Partially Autonomous Driver Assistance Systems
ISA	Intelligent Speed Adaptation
ISO	International Standards Organization
IT	Interrupting Task/Signal
ITS	Intelligent Transportation Systems
IVBSS	Integrated Vehicle-Based Safety System
IVIS	In-Vehicle Information Systems
JSD	Joint System Demonstrator
LAADS	Limited-Ability Autonomous Driving Systems
LCT	Lane Change Test
LDW	Lateral Drift Warning Subsystem
LDWS	Lane Departure Warning Systems
LED	Light-emitting Diode
LKAS	Lane Keeping Assist System

MEST	(South Korea) Ministry of Education Science and Technology
MLIT	Ministry of Land, Infrastructure, Tourism, and Transport
METI	Ministry of Economy, Trade and Industry
MKE	(South Korea) Ministry of Knowledge Economy
NASA	National Aeronautics and Space Administration
NASA-TLX	NASA Task Load Index
ND	Navigation Display
NHTSA	National Highway Traffic Safety Administration
N·m	Newton meter
OEMs	Original Equipment Manufacturers
ODEC	Operational Demand Evaluation Checklist
OT	Ongoing Task
PASAS	Predictive Airborne Separation Assurance System
PATH	California Partners for Advanced Transportation TecHnology
PRT	Personal Rapid Transit
RDCWS	Road-Departure Crash Warning System
RGT	Repertory Grid Technique
RPA	Remotely Piloted Aircraft
S&G ACC	Stop and Go ACC
SAFO	Safety Alert for Operators
SARTRE	Safe Road Trains for the Environment
Teleop	Teleoperated Uninhabited Ground Vehicles
TJA	Traffic Jam Assist
TIS-B	Traffic Information Services - Broadcast
TTC	Time-to Collision
UAs	Unnecessary Alarms
UAVs	Uninhabited Air Vehicles
UGVs	Uninhabited Ground Vehicles
UR:BAN	Urban Space: User-Oriented Assistance Systems and Network
USDOT	US Departmet nt of Transportation
UV	Unmanned Vehicle
VDU	Visual Display Unit
V2I	Vehicle to Infrastructure
V2V	Vehicle to Vehicle
WLAN	Wireless Local Area Networking

Appendix B. Study Summary Tables

Table B-1. Summary of Key Relevant European Commission Studies

Relevant Automation Level	Study Type	Summary
L3	Simulator, Test Vehicles, Demonstration	**Study:** Chan, 2012 **Project:** SARTRE **Key Building Blocks:** Lane Departure Prevention, FCW/CIB, Stability – Control Awareness • Investigated close-headway platooning as an early step in automation deployment. • Demonstrated a road train with 3 and 4 vehicles in up to 90 km/h at Hällered proving ground in Sweden.
L3	Simulator	**Study:** Larburu et al., 2010 **Related Project:** SARTRE **Key Building Blocks:** Lane Departure Prevention, FCW/CIB, Stability – Control Awareness • Investigated close-headway platooning as an early step in automation deployment. • Demonstrated a road train with 3 and 4 vehicles at up to 90 km/h at Hällered proving ground in Sweden.
L2	Simulator, Test Vehicle	**Project:** HAVEit **Key Building Blocks:** HMI – Driver Information, Situation – Control Awareness • Developed and validated a next-generation ADAS directed towards higher levels of automation as compared to the current state of the art.
L2	Simulator, Test Vehicle	**Study:** Hoeger et al., 2011 **Related Project:** HAVEit **Key Building Blocks:** HMI – Driver Information, Driver Monitoring, Situation – Control Awareness • Developed a Driver State Assessment Module for use in identifying drivers' need for automation and to make decisions when automation is to be up- or down-graded. • Available parameters for online driver drowsiness and driver distraction detection are dependent on the current automation level. • Techniques independent from the current level of automation include reaction time to specific events, indirect measures referring to additional in-vehicle activities, and direct driver monitoring.

Table B-1. (Continued)

Relevant Automation Level	Study Type	Summary
L2	Simulator, Test Vehicle	**Study:** Flemisch et al., 2010 **Related Project:** HAVEit **Key Building Blocks:** HMI – Driver Information, Driver Monitoring • High driver acceptance ratings indicated that driver monitoring (as part of an automated vehicle system) was not likely to be met with driver resistance.
L2, L3, L4	Simulator, Demonstrations	**Project:** CityMobil **Key Building Blocks:** Lane Departure Prevention, FCW/CIB, Reliability, Situation – Control Awareness, HMI – Driver Information • Research and development activities were conducted to identify, address, and, where possible, eliminate barriers blocking the implementation of automated transport systems.
L3, L4	Demonstration	**Study:** Ardelt et al., 2012 **Related Project:** CityMobil – HAD Project **Key Building Blocks:** Lane Departure Prevention, FCW/CIB, Reliability • Participants successfully navigated a 65-km test drive without the need for driver approval or intervention. • Safe driving behavior was displayed in all traffic situations, including mandatory and discretionary.
L2	Simulator, Subjective Questionnaires	**Study:** Toffetti et al., 2009 **Related Project:** CityMobil – eLane Study **Key Building Blocks:** Situation – Control Awareness, HMI – Driver Information. • Found that an interface that incorporates vocal messages is more effective in acquiring driver attention than non-speech interfaces (i.e., a system incorporating tones such as beeps). • Non-speech auditory messages should only be used as an alert, while more complex messages should be communicated by vocal means.
L2	Simulator	**Study:** Merat & Jamson, 2008 **Related Project:** CityMobil - Institute for Transport Studies, Leeds, United Kingdom, Study **Key Building Blocks:** Situation – Control Awareness, HMI – Driver Information • Drivers may have reduced situational awareness during automated driving and an overreliance on the automated system.

Relevant Automation Level	Study Type	Summary
		• Automated driving systems must keep drivers engaged and in the loop.
L2	Simulator	**Project:** ISi-PADAS **Key Building Blocks:** FCW/CIB, Driver Monitoring, HMI-Driver Information, Situation – Control Awareness • A tool-supported risk-based design methodology was introduced to enable evaluation of hazards associated with human error and/or inadequate driver behavior.
L2	Simulator	**Study:** Muhrer et al., 2012 **Related Project:** ISi-PADAS **Key Building Blocks:** FCW/CIB, Driver Monitoring, HMI-Driver Information, Situation – Control Awareness • FCW systems led to earlier driver reaction times. • While literature suggests that drivers' trust in FCW+ leads to an attention shift away from the driving task, researchers found no differences in eye-gaze behavior or engagement in secondary tasks when driving with the FCW+. • Driver attention was, however, diverted to the cockpit when visual HMI symbols appeared. • Partially autonomous driver assistance systems such as FCW+ are necessary as warnings alone are not enough to prevent accidents.

Table B-2. Additional International Projects Contributing to the Development of Automated Driving

Relevant Automation Level(s)	Study Type	Summary
L2, L3	Simulator	**Study:** Carsten et al., 2012 **Related Project:** EASY Project **Key Building Blocks:** Lane Departure Prevention, FCW/CIB, Driver Monitoring, HMI – Driver Information • Participants were more likely to engage in secondary tasks as levels of automation increased. • In the semi-automated scenario, drivers were more likely to engage in secondary tasks when lateral control was in place versus longitudinal control.

Table B-2. (Continued)

Relevant Automation Level(s)	Study Type	Summary
		• More research is needed to further explain difference in automation levels and between longitudinal and lateral control. • Additional research is needed to understand how driver attention and interaction with secondary tasks may change over time.
L2	Simulator	**Study:** Birrell & Young, 2011 **Related Project:** Foot-LITE **Key Building Blocks:** HMI – Driver Information, Situation – Control Awareness • A single unified display, based on EID principles, had wider benefits on speed, acceleration, driver mental workload, and distraction than dashboard display developed according to best practices (European Commission, 2008 and ISO 2574, 2004). • Dashboard display influenced braking more than the single unified display. • Future research should examine the potential use of EID-based displays within different levels of automation.
L1	Naturalistic, Track testing, Simulator	**Study:** Carsten et al., 2008 **Related Project:** ISA-UK **Key Building Blocks:** FCW/CIB, HMI – Driver Information, Situation – Control Awareness • Mandatory ISA system was more useful than the voluntary system in terms of usefulness, but there was more driver frustration associated with the mandatory system.
L2, L3	Simulator	**Study:** Brookhuis & De Waard, 2008 **Related Project:** Netherlands ADPTS Phileas Evaluation **Key Building Blocks:** HMI – Driver Information, Situation – Control Awareness, System Interoperability, Public Safety • Driver's licenses issued for conventional driving should not be applied to the operation of automated or semi-automated vehicles unconditionally, or at least not without preparation for driving in automated/semi-automated conditions. • Recommend simulator and/or on-the-job training and separate licensing procedures.
L1	FOT	**Study:** Viti et al., 2008 **Related Project:** Netherlands FOT

Relevant Automation Level(s)	Study Type	Summary
		Key Building Blocks: HMI – Driver Information, Situation – Control Awareness, Lane Departure Prevention, FCW/CIB • Drivers preferred stable speeds as opposed to stable distance headways in medium-dense traffic conditions. • Findings suggested that the current ACC systems should be viewed as safety-comfort-enhancing because, during dense traffic conditions, drivers were likely to deactivate the ACC system to instead rely on their own driving skills.
L2	Simulator, Subjective Questionnaires	**Study:** Dijksterhuis et al. (2010) **Related Project:** European Commission REFLECT Project **Key Building Blocks:** Lane Departure Prevention: Situation – Control Awareness, HMI – Driver Information • Results indicated that although not all drivers made use of the feedback information, the results indicated positive effects, particularly for the adaptive feedback condition.
L2	Simulator, Subjective Questionnaires	**Study:** Dijksterhuis et al. (2011) **Related Project:** European Commission REFLECT Project **Key Building Blocks:** Lane Departure Prevention: Situation – Control Awareness, HMI – Driver Information • Results indicated that steering demand factors influence mental effort expenditure and using multiple measures contributes to effort assessment.
L2	Simulator, Subjective Questionnaires	**Study:** Dijksterhuis et al. (2012) **Related Project:** European Commission REFLECT Project **Key Building Blocks:** Lane Departure Prevention: Situation – Control Awareness, HMI – Driver Information • When adaptive support (i.e., support triggered by performance-based indications of effort investment) was used by participants, driving behavior improved as compared to the non-adaptive (i.e., continually updated lateral position feedback shows on a HUD) and no support modes. • Adaptive support mode was preferred mainly as a warning signal; non-adaptive feedback tended to be ignored.

Table B-2. (Continued)

Relevant Automation Level(s)	Study Type	Summary
L2	Simulator, Subjective Questionnaires	**Study:** Koustanaï et al, 2012 **Related Project:** France's MATISS Project **Key Building Blocks:** FCW/CIB, HMI – Driving Information • As drivers became more familiar with the FCW, the driver-system interactions became more effective as demonstrated through the lack of collisions, longer time headways, and better reaction times in most situations. • Although familiarization increased drivers' trust in the FCW, it did not raise system acceptance. • System did not eliminate potentially risky behaviors. • Practicing on a simulator could help drivers learn how to properly use a driver assistance system.
L3	Roadway Testing	**Study:** Broggi et al., 2012; Bertozzi et al., 2011 **Related Project:** VisLab Intercontinental Autonomous Challenge **Key Building Blocks:** FCW/CIB, Lane Departure Prevention, Situation – Control Awareness, DSRC for Connectivity, Radar-Lidar-Camera for Crash Avoidance, Sustainability, GPS for Positioning • An autonomous vehicle traversed an approximately 13,000 km route from Rome to Shanghai over a wide variety or road types, traffic situations, and weather conditions. • Overall approach to automation keyed on low-cost and highly integrated sensors. • Approximately 50 terabytes of data were collected during the expedition, which offers a resource for further algorithm development.
L3	Pilot and Preliminary Track Testing; Simulator; Questionnaires; Demonstration	**Study:** Zlocki, 2012 **Related Project:** KONVOI **Key Building Blocks:** FCW/CIB, Lane Departure Prevention, Situation – Control Awareness, Sustainability, Public Safety • Project explored the practical use of truck platoons in road freight transport. • Demonstrated a platoon of four trucks, with the first truck driven manually by a driver supported by ACC and LDWS. • Findings suggested the safe operation of platoons with no amplifications of minor interference and no significant influence on surrounding traffic. • Public indicated general acceptance of platoons, but additional education regarding platoons is needed.

Relevant Automation Level(s)	Study Type	Summary
		• Simulator findings suggest that driving distances were lower after 2 hours of platooning; platooning had no influence on lateral driving behavior.
L1	Track Testing, Expressway Testing	**Study:** Hirai et al., 2007 **Key Building Blocks:** Curve Speed Warning, HMI-Driver Information, DSRC for Connectivity, Radar-Lidar-Camera for Crash Avoidance • Developed a safety countermeasure for use on merge ramps. • Found that the service reduced aggressive merging. • Drivers were generally receptive to the service; service implementation was accompanied by an increase in drivers' perceptions of situational dangers.
L3	Demonstration; Market Research	**Study:** Tsugawa, 2012 **Related Project:** X **Key Building Blocks:** FCW/CIB, Lane Departure Prevention, Situation – Control Awareness, DSRC for Connectivity, Radar-Lidar-Camera for Crash Avoidance, Sustainability, Public Safety • Demonstrated platoons of three heavy trucks and also three heavy trucks and one light truck. • Successfully engaged in engaging, lane changing, and passenger car cut-in maneuvers. • Identified the need for a passive safety device when inter-vehicle gaps are small (i.e., under 4 m platooning) and an HMI that provides information regarding the status of the platoon. • Freight operators expect platooning to result in energy savings, congestion reduction, load reduction, and high company brand image. • Freight operators were less sure of potential safety benefits and workload reduction.
L2	Road Tests	**Study:** Kim & Son, 2011 **Related Project:** X **Key Building Blocks:** HMI – Driver Information • Explorations of age-related workload differences as demonstrated through the completion of five driving tasks revealed that older drivers needed more time to complete tasks than did younger drivers. • Older drivers spent 13.08 s and 15.60 s completing primarily visual and visual and manual tasks, exceeding guidelines suggesting device operation should take less than 2 s and visual task time should not exceed 15 s (Bischoff, 2007). • HMI designers need to include methods for reducing distraction for older drivers.

Table B-3. Studies of Automated Driving

Relevant Automation Level(s)	Study Type	Summary
L2, L3, L4	Demonstration	**Study:** Thiessen, 2011; Schoenherr, 2009 **Key Building Blocks:** GPS for Positioning, Radar-Lidar-Camera for Crash Avoidance, Lane Departure Prevention, FCW/CIB; Stability Control, Situation – Control Awareness HMI – Driver Information, Adaptability (i.e., Scalability) • Autonomous Mobility Appliqué System (AMAS) provides scalable autonomy in a single material solution agnostic of vehicle. • AMAS will be implemented via a vehicle-specific by-wire kit that provides the electronically controlled subsystems and interface for the autonomy kit. • Autonomy kit in conjunction with the by-wire kit provides leader/follower, waypoint navigation, and advanced convoy behaviors. • U.S. Army's Convoy Active Safety Technology (CAST) system is intended to be a low-cost automated following system for tactical wheeled vehicles. • By automating the driving function, the CAST system gave drivers increased opportunities to increase situational awareness. • Gap distance formation maintenance improved 150% with CAST; successful daylight driving at 85 km/h and blackout driving at 70 km/h.
L4	Implemented System	**Study:** Main, 2013 **Key Building Blocks:** GPS for Positioning, Radar-Lidar-Camera for Crash Avoidance, Lane Departure Prevention, FCW/CIB, Navigation • Eight automated vehicles developed by Israel's Ministry of Defense have been patrolling Israeli borders. • Vehicles use cameras, radar, and laser technology for situational awareness.
L4	Implemented System	**Study:** Rio Tinto, 2012 **Key Building Blocks:** Lane Departure Prevention, FCW/CIB, Navigation • Rio Tinto Yandicoogina mine in Western Australia operates a fleet of 10 Komatsu driverless haul trucks. • Company plans to implement 150 driverless trucks.

Relevant Automation Level(s)	Study Type	Summary
L3	Track Testing	**Study:** Lee, 2013; Oxford Mobile Robotics Group, 2013 **Key Building Blocks:** Radar-Lidar-Camera for Crash Avoidance, Situation – Control Awareness, HMI – Driver Information • Oxford RobotCar is a highly automated vehicle which is able to perform all the calculations needed to plan, control speed, and avoid obstacles. • As opposed to GPS, the car "learns" and develops situational awareness through the use of information obtained during prior experience (training), prior knowledge (aerial images, road plans, semantics), and automatically generated web queries. • Technology presents a low-cost alternative for the future of automated vehicle development.
L3	Simulator	**Study:** Neubauer et al, 2011 **Key Building Blocks:** Driver Monitoring • Voluntary uses of automation failed to alleviate subjective ratings of stress and fatigue states. • Voluntary use of automation failed to improve driver performance. • Similar to required automation, voluntary automation use appears to pose similar dangers to driver alertness, task engagement, and fatigue.

Table B-4. Research Approaches Supporting Automated Driving – IVIS and Associated DVI Issues

Relevant Automation Level	Study Type	Summary
L2	Simulator	**Study:** Rauch, Gradenegger, & Krüger, 2010 **Key Building Blocks:** Situation – Control Awareness; HMI – Driver Information • Drivers were able to adapt their secondary task behavior to situational demands. • Participants anticipated potential conflicts resulted in secondary task rejection or delay in critical situations). • Situational assessment prior to the start of a secondary task and adequate monitoring of situational developments during task execution are relevant processes for situational awareness in this context.
L2	Simulator	**Study:** Davidse et al., 2009 **Key Building Blocks:** Situation – Control Awareness; HMI – Driver Information

Table B-4. (Continued)

Relevant Automation Level	Study Type	Summary
		• Messages provided by a dedicated driver support system increased aspects of safety performance for younger and older drivers. • None of the messages reduced workload; some increased workload. • Increased workload may have been the result of the task added to the driving task; researchers expected this to be a temporary increase. • ADAS settings should be adjustable to car owner's general driving behavior.
L2	Simulator	**Study:** Merat et al., 2012 **Key Building Blocks:** Situation – Control Awareness; Driver Monitoring • In the absence of a secondary task, drivers' responses to critical incidents were similar in manual and highly automated conditions. • Driver performances in the driving and secondary tasks were found to be most impaired when the two were required together, especially when drivers had to resume control after a period of under-load imposed by vehicle automation.
L2	Simulator	**Study:** Kaber et al., 2012 **Key Building Blocks:** Driver Monitoring; HMI – Driver Information • Visual and cognitive distractions have independent and combined effects on driver performance, visual behavior, and workload. • Effects vary based on the level of driving control, including operational and tactile. • Visual distraction appeared to increase driver workload through more complex gaze behavior. Drivers compensated by increasing headway times; however, drivers may fail to adapt when task demands are high. • Tactile control was more sensitive to driver distraction and less conducive to adaptation. • Findings may be limited due to use of young drivers, who are more vulnerable to the influence of distraction; a broader sample population should be investigated.
L2	Simulator	**Study:** Benedetto et al., 2011 **Key Building Blocks:** Driver Monitoring; HMI – Driver Information

Relevant Automation Level	Study Type	Summary
		• Eye-blink duration showed a Gaussian-like distribution in single-task conditions, while the distribution shifted to the left of the curve in dual-task conditions. Blink length inhibition may have occurred to avoid visual information loss. • Blink rates reflected visual workload and time on task: shorter blink rates were associated with IVIS interaction during driving while long blink rates occurred with greater frequency as time spent driving increased.
L2	Simulator	**Study:** Spießl, 2011 **Key Building Blocks:** Driver Monitoring • Participants showed the greatest deviation from the lane at the first occurring automation error; the more automation errors a driver experienced, the smaller the deviation until a reaction. • Direction of road curves had an effect on the point in time when an error became recognizable as such; in right curves, the deviation from driving lane towards oncoming lane was noticed significantly earlier. • When using the ALCT, it is recommended that researchers focus on the degree of interaction and modality when selecting tasks for evaluation in an in-car dual-task scenario (versus interruptibility and information encoding). • When looking at secondary tasks, tasks involving active engagement and visual attention away from the driving scene showed the highest negative influence on driving performance • The higher the degree of realism of the driving task, the lower the influence of secondary tasks; haptic steering wheel feedback (i.e., the steering wheel action was automated) contributed to the mitigation of negative effects and helped to prioritize the primary task. • A prospective driving path display (Magic Carpet) also helped to keep the secondary task secondary and involved drivers more in their primary task for safe driving.
L2	Simulator	**Study:** Spießl & Hussmann, 2011 **Key Building Blocks:** Driver Monitoring • When driving with haptic steering wheel feedback (i.e., the steering wheel action was automated), participants required less time to redirect their attention towards the road after an automation error. • Use of haptic feedback kept drivers more engaged in the driving tasks.
L2	Simulator, On- Road trials	**Study:** Fletcher & Zelinsky, 2009 **Key Building Blocks:** Situation – Control Awareness; Driver Monitoring

Table B-4. (Continued)

Relevant Automation Level	Study Type	Summary
		• Use of driver eye gaze combined with road events to estimate drivers' observations is feasible; this system can identify events that were almost certainly missed by the driver. • Benefit of driver observation monitoring was demonstrated to suppress redundant warnings and cancel warnings "with a glance."
L2	Meta-Analysis	**Study:** Lu, Wickens, Sarter, and Sebok (2011) **Key Building Blocks:** HMI – Driver Information • Tactile interruptions are responded to 6% faster than auditory interruptions. • No statistically significant difference in accuracy between auditory and tactile task performance and no speed-accuracy tradeoff. • Tactile cues are recommended for low-complexity IT conditions and for notification alerts. • Auditory cues are recommended for high-complexity IT conditions and for urgent alerts.
L2	Meta-Analysis	**Study:** Lu et al., 2013 **Key Building Blocks:** HMI – Driver Information • Significant differences between auditory and tactile-interrupting tasks were observed as a function of moderator variables' complexity and urgency. • Audition, rather than vision, should be used for spatial and non-urgent tasks when accuracy is the primary concern and for categorical tasks when the response time is the issue of importance • Redundant auditory-visual combinations should be used for communication tasks under high workload, for alerting and tracking tasks in low workload, and when there is a small visual angle of separation.
		Study: Abbink, Mulder, & Boer, 2012 **Key Building Blocks**: HMI – Driver Information • Identified the benefits of using haptic shared control as a DVI. • Noted that the short-term benefits of haptic shared control found in the literature (e.g., faster and more accurate vehicle control; lower levels of control effort; reduced demand for visual attention). • Concluded that long-term use should be investigated. Potential areas to investigate include trust, overreliance, dependency on the system, and retention of skills.
L2	Simulator	**Study:** Mulder et al., 2012 **Key Building Blocks:** HMI – Driver Information; Curve Speed Warning

Relevant Automation Level	Study Type	Summary
		• Identified the benefits of using haptic shared control versus manual control for the negotiation of curves; less control activity was needed for realizing an improved safety performance when haptic shared control was implemented versus manual control. • Haptic shared control kept the driver in the loop, with enhanced performance at reduced control activity; this mitigates issues associated with full automation.
L2	Simulator	**Study:** Abbink & Mulder, 2010 **Key Building Blocks:** HMI – Driver Information • Noted the benefits of shared control; i.e., an intelligently designed DVI that continually shares the control authority with the human controller.
L2	Simulator	**Study:** Tsoi et al., 2010 **Key Building Blocks:** HMI – Driver Information; Lane Departure Prevention • Objective measures and subjective responses demonstrated that haptic guidance was beneficial during lane-keeping tasks; haptic feedback resulted in small, but significantly increased, performance with smoother and reduced steering activity. • Shared control haptic system provided drivers with the benefits of haptic guidance for lane-keeping while also allowing drivers to smoothly change lanes.
L2	Meta-analysis, theoretical design	**Study:** Abbink et al.,2008 **Key Building Blocks:** HMI – Driver Information; FCW/CIB • Impact of ADAS on car-following behavior can be viewed from a closed-loop perspective (i.e., drivers need to have informative feedback on the separation states they are controlling). • ADAS opportunities are the result of issues associated with automation (e.g., overreliance, loss of attention and skills) and binary warning systems (e.g., false alarms, nuisance). • Current ADAS systems have not yet addressed the communication of criticality level for tactical tasks or support of control operations at the operational level. • To address this gap, a car-following support design was proposed that provides continuous haptic feedback directly on the gas pedal. • More research is necessary to investigate potential long-term drawbacks of the haptic feedback, such as unwanted behavioral adaptation.

Table B-4. (Continued)

Relevant Automation Level	Study Type	Summary
L2	Simulator	**Study:** Mulder et al., 2008 **Key Building Blocks:** HMI – Driver Information; Curve Speed Warning • Found that the use of haptic guidance by young, inexperienced drivers resulted in reduced and smoother steering behavior. • Curve negotiation performance was improved with less steering activity compared to driving the same track without haptic guidance. • Contrary to findings in research on longitudinal haptic guidance, the standard deviation of the steering forces increased with haptic guidance forces present, which indicated a mismatch between drivers' desired steering actions and those of the guidance system; however, subjects still reported to appreciate the guidance. • To improve the guidance system and human curve negotiation behavior, future research is needed that focuses on a better matching of guidance forces to natural driving behavior.
L2	Simulator	**Study:** Abbink & Mulder, 2009 **Key Building Blocks:** HMI – Driver Information; FCW/CIB • Haptic feedback as a means of continuous guidance during manual control tasks can improve task performance while keeping the driver in the direct manual control loop.
L2	Simulator	**Study:** Mulder & Abbink, 2010 **Key Building Blocks:** HMI – Driver Information • Found that although the haptic guidance system for curve negotiation resulted in a small increase in curve negotiation performance with less control activity, the haptic guidance system resulted in a relatively large increase in steering forces, which could potentially be a disadvantage for elderly drivers.
L2	Simulator	**Study:** Mulder et al., 2011 **Key Building Blocks:** HMI – Driver Information • In the case of a properly functioning haptic shared control system, drivers decreased their obstacle hit rate compared to manual control (from 21.2% to 15.2%) when the obstacle appeared at a time-to-contact of 1.4 s. • Under faulty conditions (i.e., late activation) of the haptic shared control system, hit rate increased to 64.7% (which would be 100% in full automation).

Relevant Automation Level	Study Type	Summary
		• Haptic shared control allowed drivers to understand the fault in the system and to respond in 35% of the cases.
L2	Simulator	**Study:** Penna et al., 2010 **Key Building Blocks:** HMI – Driver Information; FCW/CIB • Applied haptic feedback (i.e., torque feedback, stiffness feedback) to a steering wheel interface and examined collision avoidance maneuvers. • Study found that haptic steering wheel feedback effectively reduced the number of crashes, decreased response time by at least 100 ms while reducing the control effort and activity in the most critical situations.
L2	Simulator, Questionnaire	**Study:** Brandt et al., 2007 **Key Building Blocks:** Lane Departure Prevention; FCW/CIB; HMI – Driver Information • Developed and preliminarily tested a proof of concept for a haptic HMI combined with a novel lane-keeping and collision avoidance system wherein the driver assumes control over the longitudinal guidance while the driver and assistance work cooperatively for lateral vehicle guidance. • Haptic features are added to the hand wheel (i.e., steering wheel) torque and throttle and brake pedal forces. • Found good driver acceptance for the lane-keeping assistance while the collision avoidance system needed improvements.
L2	Simulator	**Study:** Griffiths & Gillespie, 2005 **Key Building Blocks:** Lane Departure Navigation; HMI – Driver Information • Demonstrated that the addition of automation through haptic display could improve performance on a primary task and reduce perceptual demands or free attention for a secondary task.
L2	Simulator	**Study:** Lervag et al., 2010 **Key Building Blocks:** Lane Departure Prevention • According to participants, lane departure warning was effective, but participants would be unwilling to install it on their personal vehicles (mandatory installation on new vehicles would be acceptable though). • Participants preferred tactile warnings (seat and steering wheel) over visual and audio warnings.
L2	Real-Road track tests	**Study:** Radke et al., 2013 **Key Building Blocks:** HMI – Driver Information; FCW/CIB

Table B-4. (Continued)

Relevant Automation Level	Study Type	Summary
		• Developed a single value called the Dynamics Factor which is valid for the objective measurement of human subjective dynamics perception in automated longitudinally controlled passenger vehicles. • Correlation of the Dynamics Factor with participants' subjected ratings proved the existence of the *co-driver effect*. • Dynamics Factor allows for the adjustment of the vehicle's longitudinal dynamics behavior to the driver's expectations and preferences, thus decreasing the dangers associated with driver out-of-the-loop performance problems. • Current effort explored the use of the Dynamics Factor in rural road settings; additional research is needed on its applicability to highways and urban roads.

Table B-5. Research Approaches Supporting Automated Driving – Trust in L2 and L3 Automated Vehicles

Relevant Automation Level	Study Type	Summary
L2	Questionnaires	**Study:** Verbene et al., 2012 **Key Building Blocks:** Adaptive Cruise Control; Reliability (trust) • Authors note that social trust and system trust have a similar determinant: shared goals. • ACCs that are described as sharing driving goals with the user led to more trustworthy and acceptable judgments than ACCs that do not. • Trust was found to mediate the effects of shared driving goals and automation level on the acceptability of ACCs.
L2, L3	Simulator, Questionnaire	**Study:** Weinstock et al., 2012 **Key Building Blocks:** Navigation, Trust • The more inaccurate or unreliable a system is, the less people will trust the system, and the less satisfied they will be with the system. • People are more willing to trust a system that is aesthetically pleasing than one that does not focus on aesthetics.

Relevant Automation Level	Study Type	Summary
L2	Test Vehicle, Questionnaire, Physiological Measures	**Study:** Reimer, Mehler, & Coughlin (2010) **Key Building Blocks:** Situation – Control Awareness • Use of an assistive parallel parking technology was shown to lower driver stress levels and average heart rates. • Use of a cross-traffic alert system, while not significantly reducing driver stress levels, may provide potential safety benefits as a result of increased situational awareness and a greater likelihood that drivers will stop and yield to approaching vehicles.
L2	Simulator	**Study:** Lees & Lee, 2009 **Key Building Blocks:** Trust; FCW/CIB; HMI – Driver Information • Demonstrated that the context of an alarm influences drivers' compliance with the alarm and that the alarm influences drivers' response to the driving situation. • Two types of alarms (classified as false alarms according to signal detection theory) had substantially different effects on subsequent true alarms. • False alarms diminished trust in and compliance with a collision warning system; conversely, the context associated with unnecessary alarms (i.e., nuisance alarms) fostered trust and compliance during subsequent events. • Greater emphasis should be placed on eliminating false alarms than unnecessary alarms. • Current warning descriptions based on signal detection theory need to be expanded to represent how different types of alarms affect drivers.
L2	Simulator	**Study:** Seppelt & Lee, 2007 **Key Building Blocks:** Adaptive Cruise Control; HMI – Driver Information; Situation – Control Awareness • Applied an EID to create a visual representation of ACC behavior and then evaluated the effect of automation and display on ACC reliance, brake response, and driver intervention strategies. • Drivers relied more appropriately on ACC with the EID present than when it was not. • EID promoted faster and more consistent braking responses when braking algorithm limits were exceeded, resulting in safe following distances and no collisions. • In manual control, EID display aided time headway.

Table B-5. (Continued)

Relevant Automation Level	Study Type	Summary
L2	Simulator, Questionnaire	**Study:** Seppelt, 2009 **Key Building Blocks:** Adaptive Cruise Control; HMI – Driver Information; Situation – Control Awareness • Informs the design of automation support displays including the type of feedback to provide in domains defined by uncertainty, complexity, and time intensity. • Provides an overview of constraints and display features, conditions and visual and auditory cues.
L2	Survey	**Study:** Larsson, 2010 **Key Building Blocks:** Adaptive Cruise Control; Situation – Control Awareness; Reliability; HMI – Driver Interface • Found that the more drivers identified system limitations and reclaimed control in situations that were not included in the owner's manual. • Respondents reported mode errors and a general lack of knowledge about the system, indicating that the ACC system is not as self-explanatory as believed.
L1, L2	Survey	**Study:** Larsson, 2012 **Key Building Blocks:** Adaptive Cruise Control; Situation – Control Awareness; Reliability • Found that the more drivers used ACC systems, the more aware of its limitations they became. • Posits that ACC systems may be less detrimental to driver performance than previous research suggests and that less-than-perfect systems help to keep drivers in the loop by forcing them to reclaim control from time to time.
L1, L2	Simulation, Field Test	**Study:** Shladover et al., 2011 **Key Building Blocks:** DSRC for Connectivity; CACC; Situation – Control; HMI – Driver Information; Driver Monitoring • Found that drivers were generally comfortable with following time-gaps under 1.0 s that could be provided by a CACC system. • Males preferred a shorter time gap setting (0.6 s) than females (0.7 s). • Participants indicated time-gap settings offered by ACC resembled gaps they would keep when driving manually in light-to-medium traffic; CACC systems were closer to time-gaps they would keep when driving manually in heavy traffic.

Relevant Automation Level	Study Type	Summary
L1	Field Test	**Project:** Nowakowski et al., 2010 **Key Building Blocks:** CACC; Situation – Control; HMI – Driver Information; Driver Monitoring; HMI – Driver Information • In simulation and live tests on I-80, the researchers broadcast speeds calculated to prevent traffic flow breakdowns, with promising results; variable speed limits demonstrated significant potential to prevent traffic delays. • Study results show that CACC could substantially increase highway capacity when it reaches moderate to high market penetration. Retrofitting non-CACC vehicles with inexpensive "here I am" radios could accelerate achievement of these capacity benefits. • A wireless communications system successfully coordinated a platoon of three tractor-trailer trucks traveling at 85 km/h (53 mi/h) and in varied joining and splitting maneuvers. Fuel savings were estimated at 10 to 14 percent for the following trucks.

Table B-6. Research Approaches Supporting Automated Driving – ACC and FCWS

Relevant Automation Level	Study Type	Summary
L2	Simulation	**Project:** Fitch et al., 2008 **Key Building Blocks:** FCW/CIB • Estimated the benefits of FCW systems at approximately 21%.
L2	Naturalistic	**Project:** Murray, 2009 **Key Building Blocks:** FCW/CIB • Observed Fitch et al.'s predicted benefits of FCW in real-world operation.
L2	FOT; Questionnaires; Focus Groups; Interviews	**Study:** Wilson et al., 2007 **Related Project:** RDCWS **Key Building Blocks:** Lane Departure Prevention, Curve Speed Warning, GPS for Positioning, Driver Monitoring, HMI – Driver Information • Although both LDW and the CSW had inaccuracies in their alerts, participants gave the following features favorable ratings: LDW and CWS alert timing, LDW and CWS missed alert frequency, and LDW false-positive alert frequency.

Table B-6. (Continued)

Relevant Automation Level	Study Type	Summary
		• With the RDCW activated, a 10- to 60-percent reduction in departure conflict frequency was observed at speeds over 88 km/h (approximately 55 mi/h).
L1, L2	FOT; Questionnaires; Focus Groups; Interviews	**Study:** GM, 2005 **Related Project:** ACAS FOT **Key Building Blocks:** FCW/CIB, Driver Monitoring, HMI – Driver Information • In daytime and freeway driving, the FCW and ACC subsystems reduced the incidence of tailgating as compared to manual driving. • Driver acceptance for the ACC system was high, but mixed for the FCW. • ACC reduced driver workload and stress; false alarms generated by the FCW system resulted in system credibility issues.
L1, L2	Experiment: Surrogate target, Track Test, Surprise Breaking Trials	**Study:** Keifer et al., 2013 **Related Study:** CAMP FCW **Key Building Blocks:** FCW/CIB, Driver Monitoring, HMI – Driver Information • Based on test driver intervention rates during surprise trials, the alert timing approach evaluated, coupled with a single-stage, dual-modality (auditory plus visual) FCW alert, was found to be robust, effective, and judged appropriate across the wide range of conditions evaluated. • Results from the time-to-collision (TTC) and first look visual occlusion studies suggested that, provided the driver is looking toward the lead vehicle, the driver can quickly assess TTC and make the appropriate crash avoidance maneuver under the alert timing assumptions evaluated. • "First look" method appears to be a valid, efficient, and promising method for exploring the consequences of later FCW alert timing (e.g., crash avoidance versus crash mitigation).
L1, L2	Simulator, Questionnaire	**Project:** Funke et al., 2007 **Key Building Blocks:** Driver Monitoring FCW/CIB; Driver Monitoring; Situation – Control Awareness • Stress and automation both influenced drivers' subjective distress, but the two factors did not interact.

Relevant Automation Level	Study Type	Summary
		• Driver performance indicated that vehicle automation impacted performance similarly in the stress and no-stress conditions. • Additional research examining effects of subjective stress and vehicle automation on driver performance is needed. • Participants experience higher levels of subjective distress under the stressful driving conditions (i.e., the periodic loss of control in the simulated winter drive condition) and lower levels of distress under the automated speed control conditions; however, an interaction between stress and automation level was not found. • Further, driver performance data revealed that vehicle automation impacted performance similar in the stress and no-stress conditions. • Findings suggested the potential benefits the DSI screening tool may have for driver personnel selection efforts and the need to educate drivers, especially inexperienced drivers, about the negative consequences of stress.
L1, L2	Field Operational Test, Questionnaire, Physiological Measures	**Project:** Seto et al., 2008 **Key Building Blocks:** Adaptive Cruise Control, Situation – Control Awareness, HMI – Driver Awareness • Using the system resulted in a decrease in the frequency and magnitude of deceleration by the driver. • From the characteristics of the time-to-collision distribution, frequency of closing situations were decreased, which have contributed to the reduction of brake action by the driver. • System was effective in reducing the driver's physical and mental workload in a wide variety of driving conditions.
L1, L2	Simulator	**Project:** Vollrath et al., 2011 **Key Building Blocks:** Adaptive Cruise Control, Situation – Control Awareness • With both Adaptive Cruise Control and Cruise Control, the time spent driving above the speed limit is reduced by 20–30%. • ACC contributes to better compliance with speed limits, as well as driving slower in general. • Drivers using ACC tend to adapt speed worse when driving in fog or approaching curves. • Because it was a simulator study, the participants may not have taken the study as seriously since there is little consequence to an accident in a simulator compared to actual driving.

Table B-6. (Continued)

Relevant Automation Level	Study Type	Summary
L1, L2	Simulator	**Project:** Xiong et al., 2012 **Key Building Blocks:** Adaptive Cruise Control • Safety consequences with ACC may be primarily related to trust in automation, driving style, and understanding of system operations and personalities.
L1, L2	Track testing	**Project:** Stanton et al., 2011 **Key Building Blocks:** Adaptive Cruise Control • Target (cyclists, pedestrians, etc.) detection rate (by participants) was much higher with a radar display than standard icon or flashing icon displays. • Workload was higher with the radar display than with the other two displays. • No difference was found in driver response times with different types of displays.
L1, L2	Naturalistic	**Project:** Bao et al., 2012 **Key Building Blocks:** FCW/CIB, Public Safety • The presence of warnings increases the mean time headway. • Drivers using the warning system responded to forward conflicts 15% faster.
L1, L2	Simulator, Questionnaire	**Project:** Saffarian et al., 2012 **Key Building Blocks:** FCW/CIB; Situation – Control Awareness; Driver Monitoring • Two advantages of maintaining close headway in fog are reduced perceived risk and improved lateral control. • No differences in risk perception were found between manual and automated driving when following distances were taking into account.
L1, L2	Simulator	**Project:** Lin et al., 2009 **Key Building Blocks:** Adaptive Cruise Control • With ACC on buses, gaps between vehicles should be kept above 1.6 seconds if the driver is not distracted by a secondary task and, if the driver is continuously distracted, the time-gap should be above 2.08 seconds. • More research needs to be done on determining time gaps for curves or slopes with ACC.

)

Table B-7. Research Approaches Supporting Automated Driving – Connected Vehicles

Relevant Automation Level	Study Type	Summary
L1, L2	Naturalistic	**Project:** Nowakowski et al., 2012 **Key Building Blocks:** FCW/CIB, DSRC for connectivity • Audible alerts decreased the mean peak deceleration rate relative to the baseline only during morning commutes and off-peak hours; there is no statistical significance for evening commutes.

REFERENCES

Abbink, D., Mulder, M. & Boer, E. (2012). Haptic shared control: Smoothly shifting control authority? *Cognition, Technology & Work, 14*(1), 19-28. doi: 10.1007/s10111-011-0192-5

Abbink, D. A., Boer, E. R. & Mulder, M. (2008, June). *Motivation for continuous haptic gas pedal feedback to support car following.* Paper presented at the Intelligent Vehicles Symposium, 2008 IEEE.

Accenture (2011). *Consumers in US and UK frustrated with intelligent devices that frequently crash or freeze, new Accenture survey finds* [Press release]. Retrieved from http://newsroom.accenture.com/article_display.cfm?article_id=5146

Ardelt, M., Coester, C. & Kaempchen, N. (2012). Highly automated driving on freeways in real traffic using a probabilistic framework. *Intelligent Transportation Systems, IEEE Transactions on, 13*(4), 1576-1585. doi: 10.1109/tits.2012.2196273

Association for Unmanned Vehicle Systems International (AUVSI). (2012). *Driverless car summit 2012: Conference report* (pp. 11). Arlington, VA: author.

Audi of America (2013). *Piloted driving* [Press release]. Retrieved from http://www.audiusanews.com/pressrelease/3304/149/piloted-driving

Aviation Safety Network. (2013a). *Accident description: American Airlines N651AA.* Retrieved April 26, 2013 from http://aviation-safety.net/database/record. php?id= 19951220-1.

Aviation Safety Network. (2013b). *Accident description: Air France F-GFKC.* Retrieved April 26, 2013 from http://aviation-safety.net/database/record.php?id=19880626-0.

Balfe, N., Wilson, J. R., Sharples, S. & Clarke, T. (2011). Development of design principles for automated systems in transport control. *Ergonomics, 55*(1), 37-54. doi: 10.1080/00140139.2011.636456

*Miller, C. A. & Parasuraman, R. (2007). Designing for flexible interfaces for supervisory control. *Human Factors, 49*(1), 57-75.

Bao, S., LeBlanc, D., Sayer, J. & Flannagan, C. (2012). Heavy-truck drivers' following behavior with intervention of an integrated, in-vehicle crash warning system: A field evaluation. *Human Factors: The Journal of the Human Factors and Ergonomics Society, 54*(687). doi: 10.1177/0018720812439412

Bartels, A. (2013, March). *Automated driving - state of the art and future challenges.* Presented at the iMobility Forum Automation Working Group Meeting, Brussels. Retrieved from http://icarsupport.eu/assets/Events/20130307-WS-on-Automation-Diamant-center-Brussels/3.a-Highway-Automation.pdf

Becker, J. (2013, June). *Toward fully automated driving.* Presented at AUVSI 2013 Driverless Car Summit, Detroit, MI.

Beene, R. (2013, January). Panel sees autonomous driving, in select situations, by 2025. *Automotive News.* Retrieved from http://www.autonews.com/apps/pbcs.dll/article?AID =/20130115/OEM09/130119868/panel-sees-autonomous-driving-in-select-situations-by-2025#axzz2Os5XnrtM

Beiker, S. & Calo, R. (2010, October). Legal aspects of autonomous driving: The need for a legal infrastructure that permits autonomous driving in public to maximize safety and consumer benefit. Retrieved from http://me.stanford.edu/groups/design/automotive/ADP /Autonomous%20Driving%20-%20Essay%202010-12%20V0.5 (final).pdf

Beller, J., Heesen, M. & Vollrath, M. (2013). Improving the driver–automation interaction: An approach using automation uncertainty. *Human Factors: The Journal of the Human Factors and Ergonomics Society*. doi: 10.1177/0018720813482327

*McGuirl, J. M. & Sarter, N. B. (2006). Supporting trust collaboration and the effective use of decision aids by presenting dynamic system confidence information. *Human Factors*, *48*, 656-665.

Benmimoun, M., Pütz, A., Aust, M. L., Faber, F., Sánchez, D., Metz, B., … Malta, L. (2012). *European Large-Scale Field Operational Tests on In-Vehicle Systems (euroFOT) Deliverable 6.1 final evaluation results*. Aachen, Germany: Ford Forschungszentrum Aachen GmbH. Retrieved from http://www.eurofot-ip.eu/en/library/deliverables/sp6_d61_final_evaluation_results.htm

Beutner, A., Pettersson, S., Seglö, F., Nilsson, A., Bartels, A., To, T. B., … Hoeger, R. (2011). *HAVEit deliverable D13.1: Optimized and validated demonstration vehicles*. Regensburg, Germany: Continental Automotive GmbH Retrieved from http://www.haveit-eu.org/LH2Uploads/ItemsContent/24/HAVEit_212154_D13.1_Public.pdf

Benedetto, S., Pedrotti, M., Minin, L., Baccino, T., Re, A. & Montanari, R. (2011). Driver workload and eye blink duration. *Transportation Research Part F-Traffic Psychology and Behaviour*, *14*(3), 199-208. doi: 10.1016/j.trf.2010.12.001

Bertozzi, M., Bombini, L., Broggi, A., Buzzoni, M., Cardarelli, E., Cattani, S., …Versari, P. (2011, June). *VIAC: An out of ordinary experiment.* Paper presented at the Intelligent Vehicles Symposium (IV), 2011 IEEE.

Bigman, D. (2013, January). Driverless cars coming to showrooms by 2020, says Nissan CEO Carlos Ghosn. *Forbes*. Retrieved from http://www.forbes.com/sites/danbigman /2013/01/14/driverless-cars-coming-to-showrooms-by-2020-says-nissan-ceo-carlos-ghosn/

Birrell, S. A. & Young, M. S. (2011). The impact of smart driving aids on driving performance and driver distraction. *Transportation Research Part F-Traffic Psychology and Behaviour*, *14*(6), 484-493. doi: 10.1016/j.trf.2011.08.004

Blair, K., Sandry, J. & Rice, S. (2012). An expansion of system wide trust theory using in-vehicle automation. *Proceedings of the Human Factors and Ergonomics Society Annual Meeting*, *56*(1), 1679-1683. doi: 10.1177/1071181312561336

BMW Group (2011, August 30). *Ready for takeover! Taking BMW automated vehicle technology to the roadway* [Press release]. Retrieved from https://www.press.bmwgroup.com/pressclub/p/pcgl/pressDetail.html?title=ready-for-takeover&outputChannelId=6&id=T0119745EN&left_menu_item=node__2374

BMW Group (2013a, February 26). Heading for Europe's motorways in a highly automated BMW: BMW Group and Continental team up on next step towards highly automated driving [Press release]. Retrieved from https://www.press.bmwgroup.com/pressclub /p/pcgl/pressDetail.html?title=heading-for-europe-s-motorways-in-a-highly-automated-bmw-bmw-group-and-continental-team-up-on-next&outputChannelId=6&id=T0137270EN&left_menu_item=node__2367

BMW Group (2012, September 19). *Safe and efficient driving in tomorrow's cities* [Press release]. Retrieved from https://www.press.bmwgroup.com/pressclub/p/pcgl/pressDetail.html?title=safe-and-efficient-driving-in-tomorrow%E2%80%99s-cities-in-the-ur-ban-research-initiative-specialists-from&outputChannelId=6&id=T0132224 EN&left_menu_item=node__2367n

BMW Group (2013b, May 19). *The new BMW 5 series* [Press release]. Retrieved from https://www.press.bmwgroup.com/pressclub/p/pcgl/pressDetail.html?title=the-new-bmw-5-series&outputChannelId=6&id=T0141378EN&left_menu_item=node__2234

Bonini, D., Dupré, C. & Granger, G. (2009). How ERASMUS can support an increase in capacity in 2020. *Proceedings of the 7th International Conference on Computing, Communications and Control Technologies: CCCT 2009*, Orlando, Florida.

Brandt, T., Sattel, T. & Bohm, M. (2007). Combining haptic human-machine interaction with predictive path planning for lane-keeping and collision avoidance systems. *Proceedings of the 2007 IEEE Intelligent Vehicles Symposium*, 582-587. doi: 10.1109/ivs. 2007.4290178

Broggi, A. (2012). *Latest challenges in autonomous vehicles: A practitioner's perspective* (SAE International Paper No. 2012-01-9018). Warrendale, PA: SAE International.

Broggi, A., Cerri, P., Felisa, M., Laghi, M. C., Mazzei, L. & Porta, P. P. (2012). The VisLab Intercontinental Autonomous Challenge: An extensive test for a platoon of intelligent vehicles. *International Journal of Vehicle Autonomous Systems*, *10*(3), 147-164. doi: 10.1504/ijvas.2012.051250

Brookhuis, K. A. & De Waard, D. (2008). *Consequences of automation for driver behaviour and acceptance.* Groningen: University of Groningen.

Bunkley, N. (2012, July 16). The car without a driver. *Automotive News*. Retrieved from http://www.autonews.com/apps/pbcs.dll/article?AID=/20120716/OEM06/307169920/1143#axzz2Os5XnrtM

Burnes C. M. & Hajdukiewicz, J. R. (2004) *Ecological interface design*. Boca Raton, FL: CRC Press.

Carsten, O., Fowkes, M., Frank, L., Chorlton, K., Jamson, S., Tate, F. & Simpkin, F. (2008). *Intelligent speed adaptation* (Final Report). The University of Leeds and MIRA Ltd.

Carsten, O., Lai, F. C. H., Barnard, Y., Jamson, A. H. & Merat, N. (2012). Control task substitution in semiautomated driving: Does it matter what aspects are automated? *Human Factors: The Journal of the Human Factors and Ergonomics Society*, *54*(5), 747-761. doi: 10.1177/0018720812460246

Chan, E. (2012, October). *An overview of the SARTRE platooning project* (Paper 2012-01-9019). Presented at SAE Convergence, Detroit, USA.

Chen, J. Y. C. & Barnes, M. J. (2012). Supervisory control of multiple robots: Effects of imperfect automation and individual differences. *Human Factors: The Journal of the Human Factors and Ergonomics Society*, *54*(2), 157-174. doi: 10.1177/0018720811435843

Chen, J. Y. C., Durlach, P. J., Sloan, J. A. & Bowens, L. D. (2008). Human robot interaction in the context of route reconnaissance missions. *Military Psychology*, *20*, 135-149. doi: 10.1080/08995600802115904

Chen, J. Y. C. & Terrence, P. I. (2009). Effects of imperfect automation and individual differences on concurrent performance of military and robotics tasks in a simulated multitasking environment. *Ergonomics*, *52*(8), 907-920. doi: 10.1080/00140130802680773

Christensen, J. C. & Estepp, J. R. (2013). Coadaptive aiding and automation enhance operator performance. *Human Factors: The Journal of the Human Factors and Ergonomics Society*. doi: 10.1177/0018720813476883

*Wilson, G. F. & Russell, C. A. (2007). Performance enhancement in an uninhabited air vehicle task using psychophysiologically determined adaptive aiding. *Human Factors*, *43*, 1005–1018.

CItyMobil. (2008). Human factors' aspects in automated and semiautomatic transport systems: State of the art (CityMobil Project Deliverable 3.2.1). Retrieved April 15, 2013, from http://www.citymobil-project.eu/site/en/documenten_deliverables.php

Coelingh, E. (2013, June) *Autonomous driving: The SARTRE project and beyond.* Presented at AUVSI 2013 Driverless Car Summit, Detroit, MI.

Cummings, M. L., Mastracchio, C., Thornburg, K. M. & Mkrtchyan, A. (2013). Boredom and distraction in multiple unmanned vehicle supervisory control. *Interacting with Computers*, *25*(1), 34-47.

Cunningham, J. (2007). Break the monotony. *Professional Engineering*, *20*(20), 33-33.

Cunningham, W. (2011, September 14). Valeo demonstrates parking by iPhone. *CNET*. Retrieved from http://reviews.cnet.com/8301-13745_7-20105958-10310244/valeo-demonstrates-parking-by-iphone/

Daimler AG (2013, February 8). *New level of safety: Intelligent assistance systems with a keener eye* [Press release]. Retrieved from: http://media.daimler.com/dcmedia/0-921-1553297-1-1570342-1-0-0-1570849-0-1-12759-614216-0-0-0-0-0-0-0.html?TS=1364849929687

Daimler AG (2012a, November 19). *New assistance systems: The helpers in the background* [Press release]. Retrieved from http://media.daimler.com/dcmedia/0-921-1549267-1-1549448-1-0-0-1549717-0-1-12759-614216-0-0-0-0-0-0-0.html?TS=1364849548507

Daimler AG (2012b, November 22). *Daimler's '6D Vision' innovation is awarded the Karl Heinz Beckurts Prize* [Press release]. Retrieved from http://media.daimler.com/dcmedia/0-921-656548-1-1552158-1-0-0-0-0-0-11817-0-0-1-0-0-0-0-0.html

Davidse, R. J., Hagenzieker, M. P., van Wolffelaar, P. C. & Brouwer, W. H. (2009). Effects of in-car support on mental workload and driving performance of older drivers. *Human Factors: The Journal of the Human Factors and Ergonomics Society*, *51*(4), 463-476. doi: 10.1177/0018720809344977

DARPA (2013). Urban Challenge. Retrieved from http://archive.darpa.mil/grandchallenge/

Dehais, F., Causse, M. & Tremblay, S. (2011). Mitigation of conflicts with automation: Use of cognitive countermeasures. *Human Factors: The Journal of the Human Factors and Ergonomics Society*, *53*(5), 448-460. doi: 10.1177/0018720811418635

Dijksterhuis, C., Kroiβ, C. & Waard, D. D. (2010). Adaptive driving support: Information about the vehicle's lateral position. *Human Factors: A System View of Human, Technology, and Organization* (Annual Conference of the European Chapter of the Human Factors and Ergonomics Society 2009), 71-87.

*Ven Der Laan, J. D., Heino, A. & De Waard, D. (1997). A simple procedure for the assessment of acceptance of advanced transport telematics. *Transportation Research – Part C: Emerging Technologies*, *5*, 1-10.

*Zijlstra, F. R. H. (1993). *Efficiency in work behavior. A design approach for modern tools.* Ph.D. Thesis. Delft University of Technology. Delft, The Netherlands: Delft University Press.

Dijksterhuis, C., Brookhuis, K. A. & De Waard, D. (2011). Effects of steering demand on lane keeping behaviour, self-reports, and physiology: A simulator study. *Accident Analysis & Prevention*, *43*(3), 1074-1081. doi: http://dx.doi.org/10.1016/j.aap.2010.12.014

Dijksterhuis, C., Stuiver, A., Mulder, B., Brookhuis, K. A. & De Waard, D. (2012). An adaptive driver support system: User experiences and driving performance in a simulator. *Human Factors: The Journal of the Human Factors and Ergonomics Society*, *54*(5), 772-785. doi: 10.1177/0018720811430502

Duxbury, C. & Stoll, J. (2012, December 3). Volvo Stakes Its Claim to Driverless Vehicles. *The Wall Street Journal.* Retrieved from http://online.wsj.com/article/SB1000142412788732375110457814735267389 2548.html

Endsley, M. (1987). The application of human factors to the development of expert systems for advanced cockpits. *Proceedings of the Human Factors Society*, *31*, 1388-1392.

Endsley, M. R. & Kaber, D. B. (1999). Level of automation effects on performance, situation awareness and workload in a dynamic control task. *Ergonomics*, *42*(3), 462-492.

European Commission (2008). Commission Recommendation on safe and efficient in-vehicle information and communication systems: Update of the European Statement of Principles on Human-Machine Interface (Report No. 2008/653/EC). *Official Journal of the European Union.* Brussels, Belgium.

Federal Aviation Administration (FAA). (2013, January). *Manual flight operations* (SAFO 13002). Washington, DC: FAA.

Feigh, K. M., Dorneich, M. C. & Hayes, C. C. (2012). Toward a characterization of adaptive systems: A framework for researchers and system designers. *Human Factors: The Journal of the Human Factors and Ergonomics Society*, *54*(6), 1008-1024. doi: 10.1177/0018720812443983

Finn, A. & Scheding, S. (2010). Developments and challenges for autonomous unmanned vehicles: A compendium (Vol. 3). Springer

Fitch, G.M., Rakha, H.A., Arafeh, M., Blanco, M., Gupta, S.K., Zimmerman, R.P. & Hanowski, R.J. (2008). *Safety benefit evaluation of a forward collision warning system* (Report No. DOT HS 810 910). Washington, DC: National Highway Traffic Safety Administration.

Fitts, P.M. (Ed.). (1951). *Human engineering for an effective air navigation and traffic control system.* Washington, DC: National Research Council.

Flemisch, F., Kaussner, A., Petermann, I., Schieben, A. & Schömig, N. (2010). *HAVEit: Deliverable D33.6 validation of concept on optimum task repartition.* Regensburg, Germany: Continental Automotive GmbH Retrieved from http://www.haveit-eu.org/LH2Uploads/ItemsContent/24/HAVEit_212154_D33.6_Public.pdf.

Fletcher, L. & Zelinsky, A. (2009). Driver inattention detection based on eye gaze-road event correlation. *International Journal of Robotics Research*, *28*(6), 774-801. doi: 10.1177/0278364908099459

Ford Motor Company (2012a, June 26). *Ford develops 'Traffic Jam Assist' and new parking technology to help address future mobility challenges* [Press release]. Retrieved from http://media.ford.com/article_display.cfm?article_id=36719

Ford Motor Company (2012b, June 27). *Ford research developing intelligent system to help drivers manage stressful situations on the road* [Press release]. Retrieved from http://media.ford.com/article_display.cfm?article_id=36728

Funke, G., Matthews, G., Warm, J. S. & Emo, A. K. (2007). Vehicle automation: A remedy for driver stress? *Ergonomics*, *50*(8), 1302-1323. doi: 10.1080/00140130701318830

Gasser, T. M. & Westhoff, D. (2012). *BASt-study: Definitions of automation and legal issues in Germany*. Presented at the TRB Road Vehicle Automation Workshop, Irvine, CA.

General Motors Corporation (2012, April 20). *Self-driving cars in Cadillac's future* [Press release]. Retrieved from http://media.gm.com/media /us/en/cadillac/news.detail.html/ content/Pages/news/us/en/2012/Apr/0420_cadillac.html

General Motors Corporation (2013, April 29). *'Super Cruise' takes on real-world traffic scenarios: Cadillac's semi-automated vehicle technology undergoes further testing* [Press release]. Retrieved from http://media.gm.com/content/media/us/en/cadillac/news.detail. html/content/Pages/news/us/en/2013/Apr/0429-cadillac-super-cruise.html

General Motors Corporation. (2005). *Automotive collision avoidance system field operational test (ACAS FOT) final program report* (Report No. DOT HS 809 886). Washington, DC: National Highway Traffic Safety Administration.

Green, P., Levison, W., Paelke, G. & Serafin, C. (1994). *Preliminary human factors design guidelines for driver information systems* (Report No. FHWA-RD-94-087). Washington, DC: Federal highway Administration.

Griffiths, P. G. & Gillespie, R. B. (2005). Sharing control between humans and automation using haptic interface: Primary and secondary task performance benefits. *Human Factors*, *47*(3), 574–590.

Guznov, S., Matthews, G., Funke, G. & Dukes, A. (2011). Use of the roboflag synthetic task environment to investigate workload and stress responses in UAV operation. *Behavior Research Methods*, *43*(3), 771-780. doi: 10.3758/s13428-011-0085-9

Hachman, M. (2012, March). Ford taking a slow road to self-driving cars. *PC Magazine*. Retrieved from http://www.pcmag.com/article2/0,2817,2401168,00.asp

Hansen, P. (2011) Google's self-driving cars zoom in on the future. *Hansen Report on Automotive Electronics*, *24*(6), 1-3.

Hart, S. & Staveland, L. (1988). Development of NASA-TLX (Task Load Index): Results of empirical and theoretical research. In: P. Hancock and N. Meshkati, eds. *Human mental workload* [online]. Amsterdam: North Holland, 139–183. Available from: http://ntrs.nasa.gov/archive/nasa/casi.ntrs.nasa.gov/20000004342_1999205624.pdf [Accessed 22 March 2013].

HAVEit. (2012). About HAVEit, Retrieved April 15, 2013 from http://www.haveit-eu.org/displayITM1.asp?ITMID=6&LANG=EN

Hirai, S., Hatakenaka, S., Hirasawa, T., Aya, T., Nishii, S. & Nomura, K. (2007). *AHS safety service utilizing an ITS on-board unit for driving support in merging sections (Paper No. 3217)*. Paper presented at the 2007 ITS World Congress, Beijing, China. http://www.nilim.go.jp/japanese/its/3paper/pdf/071128ITSWC_2.pdf

Hoeger, R., Zeng, H., Hoess, A., Kranz, T., Boverie, S., Strauss, M., … Nilsson, A. (2011). *HAVEit: Deliverable D61.1 final report*. Regensburg, Germany Continental Automotive GmbH Retrieved from http://www.explinovo.com/public/download/documents/HAVEit _212154_D61.1_Final_Report_Published.pdf.

Honda Motor Europe Ltd. (2008, April 4). *LKAS/ACC safety features on the new Accord* [Press release]. Retrieved from http://www.honda-news.eu/cz/carnewsdetail/?id=899

iCarSupport (2013, March). iMobility Forum. *iCarSupport* Retrieved from http://icarsupport.eu/esafety-forum/

Inagaki, T. (2010). Traffic systems as joint cognitive systems: Issues to be solved for realizing human-technology coagency. *Cognition, Technology & Work, 12*(2), 153-162. doi: 10.1007/s10111-010-0143-6

INDUCT (2013, March). NAVIA: A driverless 8 passenger robotized shuttle. *INDUCT.* Retrieved from http://induct-technology.com/en/products/navia

ISO 2575 (2004). *Road vehicles – Symbols for controls, indicators and tell-tales.* Geneva, Switzerland: International Organization for Standardization.

ISO 26022:2010 (2010). *Road vehicles – ergonomic aspects of transport information and control systems – Simulated lane change test to assess in-vehicle secondary task demand.* Geneva, Switzerland: International Organization for Standardization.

J. D. Power and Associates (2012, April 26). *2012 U.S. automotive emerging technologies study: Vehicle owners show willingness to spend on automotive infotainment features* [Press release]. Retrieved from http://www.jdpower.com/content/press-release/gGOwCnW/2012-u-s-automotive-emerging-technologies-study.htm

Jääskeläinen, J. (2013, March). *Automated driving from the European perspective.* Presented at the iMobility Forum Automation Working Group meeting, Brussels. Retrieved from http://icarsupport.eu/assets/Events/20130307-WS-on-Automation-Diamant-center-Brussels/1.a.1-EC.pdf

Japan Ministry of Land, Infrastructure, Transport and Tourism (MLIT) (n.d.). Advanced Safety Vehicle: Realization of Secure and Safe Traffic Society by Harmonizing Humans and Vehicles. Retrieved from http://www.mlit.go.jp/jidosha/anzen/ 01asv/ japanese/ planning5.html

Jipp, M. (2012). Individual differences and their impact on the safety and the efficiency of human-wheelchair systems. *Human Factors: The Journal of the Human Factors and Ergonomics Society, 54*(6), 1075-1086. doi: 10.1177/0018720812443826

Kaber, D. B., Liang, Y. L., Zhang, Y., Rogers, M. L. & Gangakhedkar, S. (2012). Driver performance effects of simultaneous visual and cognitive distraction and adaptation behavior. *Transportation Research Part F-Traffic Psychology and Behaviour, 15(*5), 491-501. doi: 10.1016/j.trf.2012.05.004

Kaber, D. B. & Kim, S.-h. (2011). Understanding cognitive strategy with adaptive automation in dual-task performance using computational cognitive models. *Human Factors: The Journal of Cognitive Engineering and Decision Making, 5*(3), 309-331. doi: 10.1177/1555343411416442.

Kaber, D. B. & Prinzel, L. J., III. (2006). *Adaptive and adaptable automation design: A critical review of the literature and recommendations for future research.* (NASA/TM-2006-214504). Hampton, Virginia: National Aeronautics and Space Administration Retrieved from http://ntrs.nasa.gov/search.jsp?R=20060053373.

Kalra, N., Anderson, J. & Wachs, M. (2009). *Liability and regulation of autonomous vehicle technologies* (UCB-ITS-PRR-2009-28). Richmond, CA: University of California, Berkeley Retrieved from http://www.path.berkeley.edu/PATH/Publications/ PDF/PRR/ 2009/PRR-2009-28.pdf

Karlsruhe Institute of Technology (2013). Cognitive Automobiles. Retrieved from http://his.anthropomatik.kit.edu/english/68.php

Karlsruhe Institute of Technology (2013). Team AnnieWAY. Retrieved from http://www.mrt.kit.edu/annieway/technique.html

Keane, A. (2013, February 6). Self-driving cars more Jetsons than reality for Google designers. *Bloomberg News*. Retrieved from http://www.bloomberg.com/news/2013-02-06/self-driving-cars-more-jetsons-than-reality-for-google-designers.html

Kiefer, R. J., Cassar, M. T., Flannagan, C. A., Jerome, C. J. & Palmer, M. D. (2005). *Forward collision warning requirements project: Tasks 2 and 3A final report: Surprise braking trials, time-to-collision judgments, and "first look" maneuvers under realistic rear-end crash scenarios*. (Report No. DTFH61-01-X-00014). Washington, DC: National Highway Traffic Safety Administration.

Kessler, C. & Etemad, A. (2012). *Deliverable D6.8 FOT data*. (pp. 18). Aachen GmbH: euroFOT Consortium 2012.

Khan, A. M., Bacchus, A. & Erwin, S. (2012). Policy challenges of increasing automation in driving. *IATSS Research*, *35*(2), 79-89. doi: http://dx.doi.org/10.1016/j.iatssr.2012.02.002

Kim, M. H. & Son, J. (2011). On-road assessment of in-vehicle driving workload for older drivers: Design guidelines for intelligent vehicles. *International Journal of Automotive Technology*, *12*(2), 265-272. doi: 10.1007/s12239-011-0031-y

*Bischoff, D. (2007). *Developing guidelines for managing driver workload and distraction associated with telematics devices* (NHTSA Paper 07-0082). Washington, DC: National Highway Traffic Safety Administration.

Koustanaï, A., Cavallo, V., Delhomme, P. & Mas, A. (2012). Simulator training with a forward collision warning system: Effects on driver-system interactions and driver trust. *Human Factors: The Journal of the Human Factors and Ergonomics Society*, *54*(5), 709-721. doi: 10.1177/0018720812441796

Landry, S. J. (2009). Flight deck automation. In S. Y. Nof (Ed.), *Springer handbook of automation* (pp. 1215-1239). Heidelberg: Springer.

Langan-Fox, J., Sankey, M. J. & Canty, J. M. (2009). Human factors measurement for future air traffic control systems. *Human Factors: The Journal of the Human Factors and Ergonomics Society*, *51*(5), 595-637. doi: 10.1177/0018720809355278

Larburu, M., Sánchez, J. & Rodriguez, D. J. (2010). *Safe road trains for environment: Human factors' aspects in dual mode transport systems*. Paper presented at the 17th World Congress on Intelligent Transport Systems, Busan, Korea. www.sartre-project.eu/en/publications/Documents/T_EU00337.pdf

Larsson, A. F. L. (2012). Driver usage and understanding of adaptive cruise control. *Applied Ergonomics*, *43*(3), 501-506. doi: http://dx.doi.org/10.1016/j.apergo.2011.08.005

Larsson, A. F. L. (2010). *Issues in reclaiming control from advanced driver assistance systems*. Poster presented at the European Conference on Human Centered Design for Intelligent Transport Systems, Berlin, Germany. http://www.conference2010.humanist-vce.eu/document/Proceedings/Poster_Larsson.pdf

Lee, D. (2013, February 14). A car that is able to drive itself on familiar routes has been shown off at an event at Oxford University. *BBC News*. Retrieved from http://www.bbc.co.uk/news/technology-21462360?goback=%2Egde_4731574_member_21505885

Lee, J. D. (2008). Driving attention: Cognitive engineering in designing attractions and distractions. *The Bridge*, *38*(4), 32-38.

Lee, J. D. & See, K. A. (2004). Trust in automation: Designing for appropriate reliance. *Human Factors*, *46*, 50–80.

Lee, J. D. & Seppelt, B. D. (2009). Human factors in automation design. In S. Y. Nof (Ed.), *Springer handbook of automation* (pp. 417-436). Heidelberg: Springer.

Lees, M. N. & Lee, J. D. (2007). The influence of distraction and driving context on driver response to imperfect collision warning systems. *Ergonomics*, *50*(8), 1264-1286.

*Lee, J. D., McGehee, D. V., Brown, T. L. & Reyes, M. L. (2002). Collision warning timing, driver distraction, and driver response to imminent rear-end collisions in a high-fidelity driving simulator. *Human Factors*, *44*, 314-334.

*Tan, A. & Lerner, N. (1995). *Multiple attribute evaluation of auditory warning signals for in-vehicle crash warning systems*. Washington, DC: National Highway Transportation Safety Administration.

Lefevre, P. (2013, June). *Mobility solutions INDUCT*. Presented at AUVSI 2013 Driverless Car Summit, Detroit, MI.

Lenk, C., Haberstroh, M. & Wille, M. (2011). Interaction of Human, Machine, and Environment in Automated Driving Systems. *Transportation Research Record*, 2243, 138-145. Washington, DC: Transportation Research Board of the National Academies.

Lenz, P. (2013). *Object detection for inner city traffic scenarios*. Karlsruhe Institute of Technology. Retrieved from http://www.mrt.kit.edu/res2_2993.php

Lervag, L.-E., Moen, T. & Jenssen, G. D. (2010, October). *User acceptance of HMI solutions for lane departure warning system*. Paper presented at the 2010 European Transport Conference, Glasgow, Scotland.

Li, K. Z. H., Lindenberger, U., Freund, A. M. & Baltes, P. B. (2002). Walking while memorizing: Age-related differences in compensatory behavior. *Psychological Science*, 12(3), 230−237.

Lin, T.-W., Hwang, S.-L. & Green, P. A. (2009). Effects of time-gap settings of adaptive cruise control (ACC) on driving performance and subjective acceptance in a bus driving simulator. *Safety Science*, *47*(5), 620-625. doi: http://dx.doi.org/10.1016/j.ssci.2008.08.004

Lu, S. A., Wickens, C. D., Prinet, J. C., Hutchins, S. D., Sarter, N. & Sebok, A. (2013). Supporting interruption management and multimodal interface design: Three meta-analyses of task performance as a function of interrupting task modality. *Human Factors: The Journal of the Human Factors and Ergonomics Society*. doi: 10.1177/0018720813476298

*Oviatt, S. (2003) Multimodal interfaces. In J. Jacko & A. Sears (Eds.), *The human-computer interaction handbook: Fundamentals, evolving technologies and emerging applications* (pp. 286-304). Mahway, NJ: Lawrence Erlbaum.

*Sarter, N. B. (2002). Multimodal information presentation in support of human-automation communication and coordination. In E. Salas (Ed.), *Advances in human performance and cognitive engineering research* (pp. 13.36). New York, NY: JAI Press.

Lu, S. A., Wickens, C. D., Sarter, N. B. & Sebok, A. (2011). Informing the design of multimodal displays: A meta-analysis of empirical studies comparing auditory and tactile interruptions. In *Proceedings of the Human Factors and Ergonomics Society 55th Annual Meeting*, *55*(1), 1170-1174.

Lyons, J. B. & Stokes, C. K. (2012). Human–human reliance in the context of automation. *Human Factors: The Journal of the Human Factors and Ergonomics Society*, *54*(1), 112-121. doi: 10.1177/0018720811427034

*Lyons, J. B., Stokes, C. K. & Schneider, T. R. (2011). Predictors and outcomes of trust in teams. In N. A. Stanton (Ed.), *Trust in military teams* (pp. 31–48). Surrey, UK: Ashgate.

Main, D. (2013, March 15). Autonomous cars patrol Israeli border. *LiveScience*. Retrieved from http://www.livescience.com/27949-self-driving-cars-israel.html

Martens, M. H. & Jenssen, G. D. (2012). Behavioral adaptation and acceptance. In A. Eskandarian (Ed.), *Handbook of Intelligent Vehicles* (pp. 117-138). London: Springer London.

Mattern, H. (2013, June). *Valeo's perspective on vehicle automation*. Presented at AUVSI 2013 Driverless Car Summit, Detroit, MI.

Mattes, S. (2003). The lane change task as a tool for driver distraction evaluation. In H. Strasser, H. Rausch, & H. Bubb (Eds.), *Quality of work and products in enterprises of the future* (pp. 57-60), Stuttgart: Ergonomia Verlag.

Matthews, G. & Desmond, P. A. (2002). Task-induced fatigue states and simulated driving performance. *Quarterly Journal of Experimental Psychology*, *55*, 659–686.

Matthews, G., Desmond, P. A., Joyner, L. A. & Carcary, B. (1997). A comprehensive questionnaire measure of driver stress and affect. In E. Carbonell-Vaya & J. A. Rothengatter (Eds.), *Traffic and transport psychology: Theory and application* (pp. 317-326). Amsterdam: Pergamon Press.

Matthews, G., Joyner, L., Gilliland, K., Huggins, J. & Falconer, S. (1999). Validation of a comprehensive stress state questionnaire: Towards a state big three? In I. Merville, I. J. Deary, F. DeFruyt, & F. Ostendorf (Eds.), *Personality psychology in Europe* (Vol. 7, pp. 335-350). Tilburg: Tilburg University Press.

McBride, S. E., Rogers, W. A. & Fisk, A. D. (2010, September 27 - October 1). *Do younger and older adults differentially depend on an automated system?* Paper presented at the Human Factors and Ergonomics Society 54th Annual Meeting, San Francisco, CA.

Merat, N. & Jamson, A. H. (2008). How do drivers behave in a highly automated car? In *Proceedings of the Fifth International Driving Symposium on Human Factors in Driver Assessment, Training and Vehicle Design*, Big Sky, MT, 514–521.

Merat, N., Jamson, A. H., Lai, F. C. H. & Carsten, O. (2012). Highly automated driving, secondary task performance, and driver state. *Human Factors: The Journal of the Human Factors and Ergonomics Society*, *54*(5), 762-771. doi: 10.1177/0018720812442087

Mercedes-Benz USA. (2014). *E-Class Coupe Technology*. Retrieved from http://www.mbusa.com/mercedes/vehicles/class/class-E/bodystyle-CPE#waypoint =technology

Merritt, S. M., Heimbaugh, H., LaChapell, J. & Lee, D. (2012). I trust it, but I don't know why: Effects of implicit attitudes toward automation on trust in an automated system. *Human Factors: The Journal of the Human Factors and Ergonomics Society*. doi: 10.1177/0018720812465081

Montemerlo, M., Becker, J., Bhat, S. B., Dahlkamp, H., Dolgov, D., Ettinger, S., … Thrun, S. (2008). Junior: The Stanford entry in the urban challenge. *J. Field Robot*, *25*(9), 569-597. doi: http://dx.doi.org/10.1002/rob.v25:9

Muhrer, E., Reinprecht, K. & Vollrath, M. (2012). Driving with a partially autonomous forward collision warning system: How do drivers react? *Human Factors: The Journal of the Human Factors and Ergonomics Society*, *54*(5), 698-708. doi: 10.1177/0018720812439712

Mulder, M. & Abbink, D. A. (2010). Sharing control with elderly drivers: Haptic guidance during curve negotiation. *Proceedings of the 11th IFAC/IFIP/IFORS/IEA Symposium on*

Analysis, Design, and Evaluation of Human-Machine Systems, Vol. *11*(1) (310-315). Valenciennes, France. doi: 10.3182/20100831-4-FR-2021.00055

Mulder, M., Abbink, D. A. & Boer, E. R. (2012). Sharing control with haptics: Seamless driver support from manual to automatic control. *Human Factors: The Journal of the Human Factors and Ergonomics Society*, *54*(5), 786-798. doi: 10.1177/0018720812443984

Mulder, M., Abbink, D. A. & Boer, E. R. (2008, October). The effect of haptic guidance on curve negotiation behavior of young, experienced drivers. *Proceedings of the IEEE International Conference on Systems, Man and Cybernetics*, Singapore, 804-809. doi: 10.1109/icsmc.2008.4811377

Mulder, M. & Abbink, D. A. (2011). Correct and faulty driver support from shared haptic control during evasive maneuvers. In *Proceedings of the IEEE International Conference on Systems, Man and Cybernetics* (pp. 1057–1062). Anchorage, AK: Institute of Electrical and Electronics Engineers.

Muller, J. (2013, March 21). No hands, no feet: my unnerving ride in Google's driverless car. *Forbes*. Retrieved from http://www.forbes.com/sites/joannmuller/2013/03/21/no-hands-no-feet-my-unnerving-ride-in-googles-driverless-car/

Murray, D., Shackelford, S. & Houser, A. (2009). *Analysis of the benefits and costs of forward collision warning systems for the trucking industry* (Report No FMCSA-RRT-09-021). Washington, DC: Federal Motor Carrier Safety Administration.

Nåbo, A., Anund, A., Fors, C. & Karlsson, J. G. (2013). Drivers reasoning about automated driving in the future: A focus group study (VTI Report 788). Linköping, Sweden: The Swedish National Road and Transport Research Institute (VTI).

Nakanishi, Y. J. (2009). Vehicle and road automation. In S. Y. Nof (Ed.), *Springer handbook of automation* (pp. 1165-1180). Heidelberg: Springer.

Neyedli, H. F., Hollands, J. G. & Jamieson, G. A. (2011). Beyond identity: Incorporating system reliability information into an automated combat identification system. *Human Factors: The Journal of the Human Factors and Ergonomics Society*, *53*(4), 338-355. doi: 10.1177/0018720811413767

Neubauer, C., Matthews, G., Langheim, L. & Saxby, D. (2011). Fatigue and voluntary utilization of automation in simulated driving. *Human Factors: The Journal of the Human Factors and Ergonomics Society*, *54*(5), 734-746. doi: 10.1177/ 0018720811423261

Niederée, U., Jipp, M., Teegen, U. & Vollrath, M. (2012). Effects of observability, mood states, and workload on human handling errors when monitoring aircraft automation. *Proceedings of the Human Factors and Ergonomics Society Annual Meeting*, *56*(1), 1481-1485. doi: 10.1177/1071181312561414

Nissan Motor Company (2013, January 14). *2014 Infiniti Q50 Sports Sedan unleashes inspired performance and hospitality* [Press release]. Retrieved from http://infinitinews.com/en-US/infiniti/usa/channels/us-united-states-infiniti-models-infiniti-q50/releases/2014-infiniti-q50-sports-sedan-unleashes-inspired-performance-and-hospitality?la=1&download_url=%2Fen-US%2Finfiniti%2Fusa%2Freleases%2F1c61a007-e8da-456f-a156-25f49a4fe86f%2Fdownload%2Ff3af1d4c-92c6-4371-8bee-195b02dfc8ca

Nof, S. Y. (Ed.). (2009). *Springer handbook of automation* Heidelberg: Springer.

Nothdurft, T.; Hecker, P., Ohl, S., Saust, F., Maurer, M. Reschka, A. & Bohmer, J. R. (2011). Stadtpilot: First fully autonomous test drives in urban traffic. *Proceedings of the 14th International IEEE Conference on Intelligent Transportation Systems, Washington, DC*, 919-924. doi: 10.1109/ITSC.2011.6082883

Nowakowski, C., O'Connell, J., Shladover, S. E. & Cody, D. (2010). Cooperative adaptive cruise control: Driver acceptance of following gap settings less than one second. *Proceedings of the Human Factors and Ergonomics Society Annual Meeting*, *54*(24), 2033-2037. doi: 10.1177/154193121005402403

Nowakowski, C., Shladover, S. E., Cody, D., Bu, F., O'Connell, J., Dickey, S., … Nelson, D. (2011). *Cooperative adaptive cruise control: Testing drivers' choices of following distances* (UCB-ITS-PRR-2011 01). Richmond, CA: University of California, Berkeley Retrieved from http://cdm16255.contentdm.oclc.org/cdm/ref/collection/ p266401coll4/ id/5008.

Nowakowski, C., Vizzini, D., Gupta, S. & Sengupta, R. (2012). Evaluation of real-time freeway end-of-queue alerting system to promote driver situational awareness. *Transportation Research Record: Journal of the Transportation Research Board*, *2324*(-1), 37-43. doi: 10.3141/2324-05

Ollero, A. & Castaño, Á. R. (2009). Automation of mobility and navigation. In S. Y. Nof (Ed.), *Springer handbook of automation* (pp. 279-294). Heidelberg: Springer.

Oonk, M. & Svensson, J. (2013, March). *Automation in road transport past present and future*. Presented at the iMobility Forum Automation Working Group Meeting, Brussels. Retrieved from http://icarsupport.eu/assets/Events/20130307-WS-on-Automation-Diamant-center-Brussels/1.d-state-of-the-art.pdf

Orlady, H. W. & Barnes, R. B. (1997). *A Methodology for Evaluating the Operational Suitability of Air Transport Flight Deck System Enhancements (Technical Paper 975642)*. Warrendale, PA: SAE International. doi:10.4271/975642.

O'Toole, R. (2009). *Gridlock: Why we're stuck in traffic and what to do about it.* Washington, DC: Cato Institute Press.

Oxford Mobile Robotics Group (2013). *RobotCar UK*. Retrieved April 15, 2013 from http://mrg.robots.ox.ac.uk/robotcar/

Parasuraman, R. & Davies, D. R. (1977). A taxonomic analysis of vigilance. In R.R. Mackie, (ed.) *Vigilance: Theory, operational performance and physiological correlates* (pp. 559–574). New York: Plenum.

Parasuraman, R., Cosenzo, K. A. & De Visser, E. (2009). Adaptive automation for human supervision of multiple uninhabited vehicles: Effects on change detection, situation awareness, and mental workload. *Military Psychology*, *21*(2), 270-297. doi: 10.1080/08995600902768800

Parasuraman, R. & Manzey, D. H. (2010). Complacency and bias in human use of automation: An attentional integration. *Human Factors*, *52*, 381–410.

Parasuraman, R., Sheridan, T. B. & Wickens, C. D. (2008). Situation awareness, mental workload, and trust in automation: Viable, empirically supported cognitive engineering constructs. *Journal of Cognitive Engineering and Decision Making*, *2*(2), 140-160. doi: 10.1518/155534308X284417

Parasuraman, R. & Wickens, C. D. (2008). Humans: Still vital after all these years of automation. *Human Factors: The Journal of the Human Factors and Ergonomics Society*, *50*(3), 511-520. doi: 10.1518/001872008x312198

Penna, M. D., Van Paassen, M. M., Abbink, D. A. & Mulder, M. (2010). Reducing steering wheel stiffness is beneficial in supporting evasive maneuvers. *Proceedings of the 2010 IEEE International Conference on Systems Man and Cybernetics, Istanbul, Turkey*, 1628-1635. doi. 10.1109/icsmc.2010.5642388

Pickup, L., Wilson, J. & Lowe, E. (2010). The operational demand evaluation checklist (ODEC) of workload for railway signalling. *Applied Ergonomics*, *41*(3), 393-402. doi: http://dx.doi.org/10.1016/j.apergo.2009.09.003

Radke, T., Lewandowitz, L., Gauterin, F. & Kaczmarek, T. (2013). Operationalisation of human drivers' subjective dynamics perception in automated longitudinally controlled passenger vehicles. *International Journal of Vehicle Autonomous Systems*, *11*(1), 86-109. doi: 10.1504/ijvas.2013.052275

Rauch, N., Gradenegger, B. & Kruger, H. P. (2008). User strategies for the interaction with in-vehicle devices while driving. *Intelligent Transport Systems, IET*, *2*(4), 266-275. doi: 10.1049/iet-its:20080025

Reichenbach, J., Onnasch, L. & Manzey, D. (2011). Human performance consequences of automated decision aids in states of sleep loss. *Human Factors: The Journal of the Human Factors and Ergonomics Society*, *53*(6), 717-728. doi: 10.1177/0018720811418222

*Griefahn, B., Künemund, C., Bröde, P. & Mehnert, P. (2001). Zur Validität der deutschen Übersetzung des Morningness-Eveningness-Questionnaires von Horne und Östberg [Introduction: Operator functional state in the analysis of complex performance]. *Somnologie*, 5, 71–80.

*Manzey, D., Bleil, M., Bahner-Heyne, J. E., Klostermann, A., Onnasch, L., Reichenbach, J. & Röttger, S. (2008b). Auto-CAMS 2.0 manual. Retrieved from http://www.aio.tu-berlin.de/?id=30492

Reimer, B., Mehler, B. & Coughlin, J. F. (2010). *An evaluation of driver reactions to new vehicle parking assist technologies developed to reduce driver stress*. Cambridge, MA: New England University Transportation Center, Massachusetts Institute of Technology.

Riley, V. (1989). A general model of mixed-initiative human-machine systems. *Proceedings of the Human Factors Society*, *33*, 124-128.

Rio Tinto (2012, July 19). *Rio Tinto's driverless trucks achieve a new milestone at Yandicoogina mine under Mine of the Future™ programme* [Press release]. Retrieved from http://www.riotinto.com.au/documents/120719_Rio_Tintos_driverless_ trucks_ achieve_a_new_milestone_at_Yandicoogina_mine_under_Mine_of_the_Future_programme.pdf

Rodriguez, S. (2012, November 30). L.A. Auto Show: Animal detecting systems coming as soon as 2013. *Los Angeles Times*. Retrieved from http://articles.latimes.com/2012/nov/30/autos/la-fi-hy-autos-laas-animal-detection-night-vision-autoliv-20121129

Röttger, S., Bali, K. & Manzey, D. (2009). Impact of automated decision aids on performance, operator behaviour and workload in a simulated supervisory control task. *Ergonomics*, *52*(5), 512-523. doi: 10.1080/00140130802379129

Rovira, E. & Parasuraman, R. (2010). Transitioning to future air traffic management: Effects of imperfect automation on controller attention and performance. *Human Factors: The Journal of the Human Factors and Ergonomics Society*, *52*(3), 411-425. doi: 10.1177/0018720810375692

Ruigrok, R. C. J. & Hoekstra, J. M. (2007). Human factors evaluations of free flight: Issues solved and issues remaining. *Applied Ergonomics*, *38*(4), 437-455. doi: http://dx.doi.org/10.1016/j.apergo.2007.01.006

Rupp, J. D. & King, A. G. (2010). *Autonomous driving - a practical roadmap.* Warrendale, PA: SAE International. Retrieved from http://www.fujitsu.com/downloads/MICRO/fma/marcom/convergence/data/papers/2010-01-2335.pdf

Saffarian, M., de Winter, J. C. F. & Happee, R. (2012). *Automated driving: Human-factors issues and design solutions.* Proceedings of the Human Factors and Ergonomics Society Annual Meeting, 56(1), 2296-2300. doi: 10.1177/1071181312561483

*Adell, E., Várhelyi, A. & dalla Fontana, M. (2011). The effects of a driver assistance system for safe speed and safe distance - a real life field study. *Transportation Research Part C: Emerging Technologies*, *19*(1), 145-155.

*Flemisch, F., Heesen, M., Hesse, T., Kelsch, J., Schieben, A. & Beller, J. (2012). Towards a dynamic balance between humans and automation: Authority, ability, responsibility and control in shared and cooperative situations. *Cognition, Technology & Work*, *14*, 3–18.

*Pauwelussen, J. & Feenstra, P. J. (2011). Driver Behavior analysis during ACC activation and deactivation in a real traffic environment. *IEEE Transactions on Intelligent Transportation Systems*, *11*, 329-338.

*Ran, B., Leight, S., Johnson, S. & Huang, W. (1997) Human factors studies in evaluation of automated highway system attributes. *Transportation Research Part F: Traffic Psychology and Behaviour*, *6*, 233-248.

*Stanton, N. & Young, M. (2005). Driver behavior with adaptive cruise control. Ergonomics, 48, 1294–1313.

*Rudin-Brown, C. & Parker, H. (2004). Behavioural adaptation to adaptive cruise control (ACC): implications for preventive strategies. *Transportation Research Part F: Traffic Psychology and Behaviour*, *7*, 59–76.

Saffarian, M., Happee, R. & Winter, J. C. F. d. (2012). Why do drivers maintain short headways in fog? A driving-simulator study evaluating feeling of risk and lateral control during automated and manual car following. *Ergonomics*, *55*(9), 971-985. doi: 10.1080/00140139.2012.691993

Sakai, S. (2013, March). *Introduction to developments on automation – MLIT.* Presented at the iMobility Forum Automation Working Group meeting, Brussels.

Salinger, J. (2013, June). *The road to driverless vehicles: A GM perspective.* Presented at AUVSI 2013 Driverless Car Summit, Detroit, MI.

Schmitt, P. (2013, June). *Intelligent vehicles: A transport industry perspective.* Presented at AUVSI 2013 Driverless Car Summit, Detroit, MI.

Schoenherr, E. (2009, July-September). Moving future convoy operations with convoy active safety technologies (CAST). *Accelerate Magazine*. Retrieved from http://tardec.army.mil/Documents/AM0709_Moving_Future_Convoy_Ops.pdf

Schumacher, C. (2013, June). *Continental: Automation from the supplier's perspective.* Presented at AUVSI 2013 Driverless Car Summit, Detroit, MI.

Seppelt, B. D. (2009). *Supporting operator reliance on automation through continuous feedback.* Ph.D. Dissertation, University of Iowa, Iowa City, Iowa. Retrieved from http://ir.uiowa.edu/etd/432

Seppelt, B. D. & Lee, J. D. (2007). Making adaptive cruise control (ACC) limits visible. *Int. J. Human-Computer Studies*, *65*, 195-205.

Seto, Y., Takae, Y., Sugano, T., Satao, R., Kurata, K. & Kobayashi, M. (2008). Development of distance control assist system with accelerator force control and braking force control. In *Proceedings of the 15th World Congress on Intelligent Transport Systems and ITS America's 2008 Annual Meeting.*

Sheridan, T. (1970). On how often the supervisor should sample. *IEEE Transactions on Systems Science and Cybernetics*, *SSC-6*, 140-145.

Sheridan, T. B. (2006). Next generation air transportation systems: Human-automation interaction and organizational risks. *Proceedings of the Second Symposium on Resilience Engineering*, Juan-les-Pins, France.

Sheridan, T. B. & Nadler, E. D. (2006). *A review of human-automation interaction failures and lessons learned October 2006* (DOT-VNTSC-NASA-06-01). Washington, DC: National Aeronautics and Space Administration Retrieved from http://ntl.bts.gov/lib/34000/34700/34709/DOT-VNTSC-NASA-06-01.doc

Sheridan, T. B. & Parasuraman, R. (2005). Human-automation interaction. *Reviews of Human Factors and Ergonomics*, *1*(1), 89-129. doi: 10.1518/155723405783703082

Sheridan, T. B. & Verplank, W. L. (1978). *Human and computer control of undersea teleoperators*. Arlington: Office of Naval Research.

Sherman, P. J. (1997). *Aircrews evaluation of flight deck automation training and use: Measuring and ameliorating threats to safety* (No. Technical Report 97-2). Washington, DC: Federal Aviation Administration.

Shladover, S. E. (2012a). *Literature review on recent international activity in cooperative vehicle – highway automation systems* (FHWA-HRT-13-025). Washington, DC: Federal Highway Administration Retrieved from http://www.fhwa.dot.gov/advancedresearch/pubs/13025/13025.pdf

Shladover, S. E. (2012b). *Recent international activity in cooperative vehicle-highway automation systems* (FHWA-HRT-12-033). Washington, DC: Federal Highway Administration Retrieved from http://www.fhwa.dot.gov/advancedresearch/pubs/12033/12033.pdf.

Shladover, S. E., Lu, X.-Y., Nowakowski, C. & Su, D. (2011). *Development and evaluation of selected mobility applications for vii* (CA11-2138). Sacramento, CA: California Department of Transportation

Shladover, S. E., Nowakowski, C., Cody, D., Bu, F., O'Connell, J., Spring, J., ... Nelson, D. (2009). *Effects of cooperative adaptive cruise control on traffic flow: Testing drivers' choices of following distances* (UCB-ITS-PRR-2009-23). Richmond, CA: University of California, Berkeley Retrieved from http://www.path.berkeley.edu/PATH/ Publications /PDF/PRR/2009/PRR-2009-23.pdf.

Sierhuis, M. (2013, June). *Towards fully autonomous vehicles*. Presented at AUVSI 2013 Driverless Car Summit, Detroit, MI.

SILICON BEAT (2012, September 25). Sergey Brin hopes people will be driving Google Robot Cars in "several years." *SILIOCONBEAT*. Retrieved from http://www.siliconbeat.com/2012/09/25/sergey-brin-hopes-people-will-be-driving-google-robot-cars-in-several-years/

Simonite, T. (2013, October 25). Data shows Google's robot cars are smoother, safer drivers than you or I. *MIT Technology Review*. Retrieved from http://www.technologyreview.com/news/520746/data-shows-googles-robot-cars-are-smoother-safer-drivers-than-you-or-i/

Smyth, G. (2012, June). *Gary Smyth, Executive Director for North American Science Labs: General Motors presentation and interview.* Presented at the 2012 Driverless Car Summit, Detroit, USA. Retrieved from www.auvsi.org

Soriano, B. (2013, June). *Getting on the Road: A state DMV perspective.* Presented at AUVSI 2013 Driverless Car Summit, Detroit, MI.

Spießl, W. (2011). *Assessment and support of error recognition in automated driving.* Ph.D., Ludwig-Maximilians-Universität München, München.

Spießl, W. & Hussmann, H. (2011). Assessing error recognition in automated driving. *IET Intelligent Transport Systems*, *5*(2), 103-103. doi: 10.1049/iet-its.2010.0102

Stanton, N. A., Dunoyer, A. & Leatherland, A. (2011). Detection of new in-path targets by drivers using stop & go adaptive cruise control. *Applied Ergonomics*, *42*(4), 592-601. doi: 10.1016/j.apergo.2010.08.016

Stanton, N. A. & Young, M. S. (2010). Vehicle automation and driving performance. *Ergonomics*, *41*(7), 1014-1028. doi: 10.1080/001401398186568

Strauss, M., Waldbauer, D., Berthold, T., Komar, M., Pfeiffer, J., Günther, W., … Lepke, H. (2010). *HAVEit: Deliverable D51.3 automated assistance in roadworks and congestion: 1st system functionality.* Regensburg, Germany: Continental Automotive GmbH. Retrieved from http://www.haveit-eu.org/LH2Uploads/ItemsContent/24/HAVEit_212154_D51.3_Public.pdf

Sumwalt, R.L., Thomas, R.J. & Dismukes, K. (2002). *Enhancing flight-crew monitoring skills can increase safety.* Paper presented at the 55th International Air Safety Seminar. Dublin, Ireland.

Theisen, B. (2011). Autonomous Mobility Appliqué System. *SAE International.* Retrieved from http://www.sae.org/events/training/symposia/ivs/presentations/2011/ bernard_theisen.pdf

Toffetti, A., Wilschut, E. S., Martens, M. H., Schieben, A., Rambaldini, A., Merat, N. & Flemisch, F. (2009). CityMobil human factor issues regarding highly automated vehicles on eLane. *Transportation Research Record* (2110), 1-8. doi: 10.3141/2110-01

Toyota Motor Corporation (2012, September 13). *Toyota Collaborative Safety Research Center launches seven new projects and four new partnerships with leading academic and research institutions* [Press release]. Retrieved from http://pressroom.toyota.com/releases/toyota+csrc+seven+new+projects+four+partnerships+leading+academic+research+institutions+sept2012.htm

Toyota Motor Corporation (2013a, January 8). *2013 Lexus LS advanced active safety features 2013 Consumer Electronics Show* [Press release]. Retrieved from http://pressroom.lexus.com/lexus/releases/fact+sheet+2013+ls+advanced+active+safety.htm

Toyota Motor Corporation (2013b, January 7). *Toyota Motor Corp. and Lexus showcase advanced active safety research vehicle at CES* [Press release]. Retrieved from http://pressroom.lexus.com/releases/toyota+lexus+advanced+active+safety+research+vehicle+ces+jan7.htm

Transport Research & Innovation Portal (2013). Project details ISI-PADAS. Retrieved from http://www.transport-research.info/web/projects/project_details.cfm?ID=37254

Tsoi, K. K., Mulder, M. & Abbink, D. A. (2010). Balancing safety and support: Changing lanes with a haptic lane-keeping support system. *Proceedings of the 2010 IEEE International Conference on Systems Man and Cybernetics, Istanbul, Turkey*, 1236-1243. doi: 10.1109/icsmc.2010.5642414

Tsugawa, S. (2012). *Energy ITS: What we learned and what we should learn.* Presented at the TRB Road Vehicle Automation Workshop, Irvine, CA.

UR:BAN (2013). Urban space: User-friendly Assistance Systems and Network Management. Retrieved from http://urban-online.org

Urmson, C. (2012, August 7). The self-driving car logs more miles on new wheels. *Google Official Blog.* Retrieved from http://googleblog.blogspot.com.au/2012/08/the-self-driving-car-logs-more-miles-on.html

Urmson, C., Anhalt, J., Bagnell, D., Baker, C., Bittner, R., Clark, M. N., … Ziglar, J. (2008). Autonomous driving in urban environments: Boss and the urban challenge. *Journal of Field Robotics*, *25*(8), 425-466. doi: 10.1002/rob.20255

Van den Broek, T. H. A., Netten, B. D., Hoedemaeker, M. & Ploeg, J. (2010, September). *The experimental setup of a large field operational test for cooperative driving vehicles at the A270.* Paper presented at the 13th International IEEE Conference on Intelligent Transportation Systems (ITSC), 2010.

van Dijke, J. & van Schijndel, M. (2012). CityMobil, advanced transport for the urban environment. *Transportation Research Record: Journal of the Transportation Research Board*, *2324*, 29-36. doi: 10.3141/2324-04

Verberne, F. M. F., Ham, J. & Midden, C. J. H. (2012). Trust in smart systems: Sharing driving goals and giving information to increase trustworthiness and acceptability of smart systems in cars. *Human Factors: The Journal of the Human Factors and Ergonomics Society*, *54*(5), 799-810. doi: 10.1177/0018720812443825

*Reeves, B. & Nass, C. (1996). *The media equation: How people treat computers, television, and new media like real people and places.* New York, NY: Cambridge University Press.

*Schuitema, G., Steg, L. & Forward, S. (2010). Explaining differences in acceptability before and acceptance after the implementation of a congestion charge in Stockholm. *Transportation Research Part A: Policy and Practice*, *44*, 99–109.

Vicente, K. J. (2002). Ecological interface design: Progress and challenges. *Human Factors*, *44*, 62-78.

Viti, F., Hoogendoorn, S. P., Alkim, T. P. & Bootsma, G. (2008). *Driving behavior interaction with ACC: Results from a field operational test in the Netherlands.* 2008 IEEE Intelligent Vehicles Symposium (pp. 745–750). Eindhoven, the Netherlands.

Vlacic, L. & Parent, M. (2009). Cybercars 2 – Final Report. Retrieved from http://cybercars2.paris-rocquencourt.inria.fr/publications/deliverables/cybercars2_summary_report_version8.pdf/view

Vlad, E. P. & Tatarnikov, V. (2011). R&M&A&S of communication based train control systems applied to urban rail transportation - a way to improve city sustainability. *Proceedings of the 2011 Proceedings - Annual Reliability and Maintainability Symposium (RAMS).*

Vollrath, M., Schleicher, S. & Gelau, C. (2011). The influence of cruise control and adaptive cruise control on driving behaviour – a driving simulator study. *Accident Analysis & Prevention*, *43*(3), 1134-1139. doi: http://dx.doi.org/10.1016/j.aap.2010.12.023

Volvo Car Group (2013, March 5). *Volvo Car Group reveals world-first cyclist detection with full auto brake in Geneva* [Press release]. Retrieved from https://www.media.volvocars.com/global/enhanced/en-gb/Media/Preview.aspx?mediaid =48219

Volvo Car Corporation (2012, September 7). *Volvo Car Corporation aims for leadership within autonomous driving technology* [Press release]. Retrieved from https://www.media.volvocars.com/global/enhanced/en-gb/media/preview.aspx?mediaid=45718

Walker Smith, B. (2013, May 4, 2013). Automated driving: Legislative and regulatory action. *The Center for Internet and Society.* Retrieved May 5, 2013, from http://cyberlaw.stanford.edu/wiki/index.php/Automated_Driving:_Legislative_and_Regulatory_Action

Walker Smith, B. (2012a). Automated vehicles are probably legal in the United States. *The Center for Internet and Society.* Retrieved from http://cyberlaw.stanford.edu/publications/automated-vehicles-are-probably-legal-united-states

Walker Smith, B. (2012b). My other car is a ... Robot? Defining vehicle automation. *The Center for Internet and Society.* Retrieved from http://cyberlaw.stanford. edu/blog/2012/02/my-other-car-robot-defining-vehicle-automation

Weinstock, A., Oron-Gilad, T. & Parmet, Y. (2012). The effect of system aesthetics on trust, cooperation, satisfaction and annoyance in an imperfect automated system. *Work - A Journal of Prevention Assessment & Rehabilitation, 41*, 258-265. doi: 10.3233/wor-2012-0166-258

Wiener, E. L. (1988). *Cockpit automation.* In E. L. Wiener & D. C. Nagal (Eds.), Human Factors in Aviation (pp. 433-461). San Diego. CA: Academic Press, Inc.

Wille, J. M., Saust, F. & Maurer, M. (2010a). Comprehensive treated sections in a trajectory planner for realizing autonomous driving in Braunschweig's urban traffic. *Procedures of the 2010 13th International IEEE Conference on Intelligent Transportation Systems*, 647-652.

Wille, J. M., Saust, F. & Maurer, M. (2010b). Stadtpilot: Driving autonomously on Braunschweig's inner ring road. *Procedures of the 2010 IEEE the Intelligent Vehicles Symposium*, 506-511.

Williams, M. (2012, October 3). Nissan previews self-driving car. *Computer World.* Retrieved from http://www.computerworld.com/s/article/9231978/Nissan_ previews_ self_driving_car

Wilson, B. H., Stearns, M. D., Koopmann, J. & Yang, C. Y. D. (2007). *Evaluation of a road-departure crash warning system* (Report No. DOT HS 810 854). Washington, DC: National Highway Traffic Safety Administration.

Wüst, C. (2013, February 1). Auto revolution: A promising future for self-driving cars. Spiegel. Retrieved from http://www.spiegel.de/international/business/driverless-cars-are-a-reality-but-face-acceptance-obstacles-a-880716.html

Xiong, H., Boyle, L. N., Moeckli, J., Dow, B. R. & Brown, T. L. (2012). Use patterns among early adopters of adaptive cruise control. *Human Factors: The Journal of the Human Factors and Ergonomics Society, 54*(5), 722-733. doi: 10.1177/0018720811434512

Zlocki, A. (2012). *KONVOI and interactIVe: Truck platooning and crash avoidance.* Paper presented at the TRB Road Vehicle Automation Workshop, Irvine, CA.

End Note

[1] Several State automated vehicle laws consider the person who activates the automated vehicle system to be the "driver" of the vehicle even if that person is not physically present in the vehicle. NHTSA, however, is not aware of any prototype automated vehicle systems that are capable of operating on public roads without the presence of a driver in the driver's seat who is ready to control the vehicle.

In: Human Factors in Automated Driving Scenarios
Editor: Jodie Reeves
ISBN: 978-1-63463-063-4

Chapter 2

HUMAN FACTORS EVALUATION OF LEVEL 2 AND LEVEL 3 AUTOMATED DRIVING CONCEPTS: CONCEPTS OF OPERATION*

Andrew Marinik, Richard Bishop, Vikki Fitchett, Justin F. Morgan, Tammy E. Trimble and Myra Blanco

ABSTRACT

The Concepts of Operation document evaluates the functional framework of operations for Level 2 and Level 3 automated vehicle systems. This is done by defining the varying levels of automation, the operator vehicle interactions, and system components; and further, by assessing the automation relevant parameters from a scenario-based analysis stand-point. Specific to the "Human Factors Evaluation of Level 2 and Level 3 Automated Driving Concepts" research effort, scenarios and literature are used to identify the range of near- to mid-term production-intent systems such that follow-on research topics with highest impact potential can be identified through commonalities in operational concepts.

SECTION 1. INTRODUCTION

This document defines the automated passenger vehicle functional domain of operations for the purposes of this project. The document addresses system functionality in terms of the vehicle's interaction with the driver, environment, and surrounding traffic. System operation, processes, and interactions are described on a basic level. The document is based on team expertise and available literature.

The document is limited to Level 2 (L2) and Level 3 (L3) automated vehicle operations for passenger cars. The focus includes transitions between levels.

* This is an edited, reformatted and augmented version of a National Highway Traffic Safety Administration sponsored document, DOT HS 812 044, issued July 2014.

The typical Concept of Operations (ConOps) document focuses on a specific system implementation to enable a developer organization to describe the capabilities and limitations in detail and to assess system impacts, costs, and long-term requirements (maintenance, etc.). From this the developer would define functional requirements based on the ConOps, followed by technical requirements and system design.

This document is not focused on system design and development; it instead proceeds from a different basis. It supports a process to understand the broad functional domain of operations for automated vehicles and, from that universe, select several representative system configurations for L2 and L3 automation. Therefore, a full set of parameters is described that refers to the operating environment and the system function. These parameters are then assessed in terms of relevance to potential future systems as envisioned by the broader industry. From this, several concepts are formed, and scenarios are developed to create a likely depiction of first-generation L2 and L3 vehicle automation. This document, therefore, establishes a context in which to pose an array of questions related to automation from which research questions will be chosen to guide the experimental portion of the project.

This document is designed to provide a definition of the system considered to ensure that all stakeholders have a baseline from which to consider all proposed research efforts. Additionally, this document generally follows the format of an IEEE 1362-1998 standard ConOps document. However, modifications have been made based on the research nature of this system. Among the most important aspects of this document are the operational definitions. These definitions offer a common understanding necessary for continued discussion, research, and reporting. Primary among them, for this effort, are the definitions of the National Highway Traffic Safety Administration (NHTSA) taxonomy for automation:

Level 0 (L0; No Automation): The driver is in complete and sole control of the primary vehicle controls (brake, steering, throttle, and motive power) at all times, and is solely responsible for monitoring the roadway and for safe operation of all vehicle controls. Vehicles that have certain driver support/convenience systems but do not have control authority over steering, braking, or throttle would still be considered "L0" vehicles. Examples include systems that provide only warnings (e.g., forward collision warning (FCW), lane departure warning (LDWS), blind spot monitoring) as well as systems providing automated secondary controls such as wipers, headlights, turn signals, hazard lights, etc. Although a vehicle with vehicle-to-vehicle (V2V) warning technology alone would be at this level, that technology could significantly augment, and could be necessary to fully implement, many of the technologies described below, and is capable of providing warnings in several scenarios where sensors and cameras cannot (e.g., vehicles approaching each other at intersections).

Level 1 (L1; Function-specific Automation): Automation at this level involves one or more specific control functions; if multiple functions are automated, they operate independently from each other. The driver has overall control, and is solely responsible for safe operation, but can choose to cede limited authority over a primary control (as in adaptive cruise control), the vehicle can automatically assume limited authority over a primary control (as in electronic stability control), or the automated system can provide added control to aid the driver in certain normal driving or crash-imminent situations (e.g., dynamic brake support in emergencies). The vehicle may have multiple capabilities combining individual driver support and crash avoidance technologies, but does not replace driver vigilance and does not assume driving responsibility from the driver. The vehicle's automated system may assist or augment the driver in operating one of the primary controls – either steering or braking/throttle controls (but not both). As a result, there is no combination of vehicle control systems working in unison that enables the driver to be disengaged from physically operating the vehicle by having his or her hands off the steering wheel AND feet off the pedals at the same time. Examples of function-specific automation systems include: cruise control, automatic braking, and lane keeping.

Level 2 (L2; Combined Function Automation): This level involves automation of at least two primary control functions designed to work in unison to relieve the driver of control of those functions. Vehicles at this level of automation can utilize shared authority when the driver cedes active primary control in certain limited driving situations. The driver is still responsible for monitoring the roadway and safe operation and is expected to be available for control at all times and on short notice. The system can relinquish control with no advance warning and the driver must be ready to control the vehicle safely. An example of combined functions enabling an L2 system is adaptive cruise control in combination with lane centering. The major distinction between L1 and L2 is that, at L2 in the specific operating conditions for which the system is designed, an automated operating mode is enabled such that the driver is disengaged from physically operating the vehicle by having his or her hands off the steering wheel AND foot off pedal at the same time.

Level 3 (L3; Limited Self-Driving Automation): Vehicles at this level of automation enable the driver to cede full control of all safety-critical functions under certain traffic or environmental conditions and in those conditions to rely heavily on the vehicle to monitor for changes in those conditions requiring transition back to driver control. The driver is expected to be available for occasional control, but with sufficiently comfortable transition time. The vehicle is designed to ensure safe operation during the automated driving mode. An example would be an automated or self-driving car that can determine when the system is no longer able to support automation, such as from an oncoming construction area, and then signals to the driver to reengage in the driving task, providing the driver with an appropriate amount of transition time to safely regain manual control. The major distinction between L2 and L3 is that at L3, the vehicle is designed so that the driver is not expected to constantly monitor the roadway while driving.

Level 4 (L4; Full Self-Driving Automation): The vehicle is designed to perform all safety-critical driving functions and monitor roadway conditions for an entire trip. Such a design anticipates that the driver* will provide destination or navigation input, but is not expected to be available for control at any time during the trip. This includes both occupied and unoccupied vehicles. By design, safe operation rests solely on the automated vehicle system.

SECTION 2. REFERENCED DOCUMENTS

[1] Trimble, T. E., Bishop, R., Morgan, J. F. & Blanco, M. (in press). *Human factors evaluation of level 2 and level 3 automated driving concepts: Past research, state of*

* Several State automated vehicle laws consider the person who activates the automated vehicle system to be the "driver" of the vehicle even if that person is not physically present in the vehicle. NHTSA, however, is not aware of any prototype automated vehicle systems that are capable of operating on public roads without the presence of a driver in the driver's seat who is ready to control the vehicle.

automation technology, and emerging system concepts. Washington, DC: National Highway Traffic Safety Administration.

[2] Dingus, T. A., Klauer, S. G., Neale, V. L., Petersen, A., Lee, S. E., Sudweeks, J. … & Knipling, R. R. (2006). The 100-car naturalistic driving study, Phase II - results of the 100-car field experiment (Report No. DOT HS 810 593). Washington, DC: National Highway Traffic Safety Administration.

[3] Najm, W. G., Koopmann, J., Smith, J. D. & Brewer, J. (2010). *Frequency of Target Crashes for IntelliDrive Safety Systems*: U.S. Department of Transportation, Research and Innovative Technology Administration, John A. Volpe National Transportation Systems Center.

[4] Blanco, M., Hickman, J. S., Olson, R. L., Bocanegra, J. L., Hanowski, R. J., Nakata, A. . . . & Bowman, D. (2008). *Investigating critical incidents, driver restart period, sleep quantity, and crash countermeasures in commercial vehicle operations using naturalistic data collection* (pp. 549). Blacksburg, VA: Virginia Tech Transportation Institute.

[5] Hickman, J. S., Knipling, R. R., Olson, R. L., Fumero, M. C., Blanco, M. & Hanowski, R. J. (2005). *Phase 1 - preliminary analysis of data collected in the drowsy driver warning system field operational test: Task 5, preliminary analysis of data collected in the drowsy driver warning system field operational test data heavy vehicle-light vehicle interaction data collection and countermeasure research project.* Blacksburg, Virginia: Virginia Tech Transportation Institute.

[6] LeBlanc, D., Buonarosa, M. L., Blankespoor, A. & Sayer, J. (2009). *Integrated vehicle-based safety systems (IVBSS) light vehicle extended pilot test summary report* (Report No. UMTRI-2009-13). Ann Arbor, MI: University of Michigan Transportation Research Institute.

[7] Daimler AG (2013, February). *New level of safety: Intelligent assistance systems with a keener eye* [Press release]. Retrieved from http://media.daimler.com/dcmedia/0-921-1553297-1-1570342-1-0-0-1570849-0-1-12759-614216-0-0-0-0-0-0-0.html?TS=1364849929687

[8] Beutner, A., Pettersson, S., Seglö, F., Nilsson, A., Bartels, A., Binh To, T. … & Hoess, A. (2011). *HAVEit deliverable 13.1 optimized and validated demonstration vehicles.* Retrieved from http://www.haveit-eu.org/LH2Uploads/ ItemsContent/24/ HAVEit_212154_D13.1_Public.pdf

[9] Mercedes-Benz USA. (2014). *E-Class Coupe Technology.* Retrieved from http://www.mbusa.com/mercedes/vehicles/class/class-E/bodystyle-CPE#waypoint =technology

[10] General Motors Corporation (2012). *Self-driving cars in Cadillac's future* [Press release]. Retrieved from http://media.gm.com/media/us/en/ cadillac/news.detail.html/ content/Pages/news/us/en/2012/Apr/0 420_cadillac.html

[11] Nissan Motor Company (2013, January). *2014 Infiniti Q50 Sports Sedan unleashes inspired performance and hospitality* [Press release]. Retrieved from http://infinitinews.com/en-US/infiniti/usa/channels/us-united-states-infiniti-models-infiniti-q50/releases/2014-infiniti-q50-sports-sedan-unleashes-inspired-performance-and-hospitality?la=1&download_url=%2Fen-US%2Finfiniti%2Fusa%2Freleases%2F1c61a007-e8da-456f-a156-25f49a4fe86f%2Fdownload%2Ff3af1d4c-92c6-4371-8bee-195b02dfc8ca

[12] BMW Group (2011, August). *Ready for takeover! Taking BMW automated vehicle technology to the roadway* [Press release]. Retrieved from https://www.press.bmwgroup.com/pressclub/p/pcgl/pressDetail.html?title=ready-for-takeover&outputChannelId=6&id=T0119745EN&left_menu_item=node__2374

[13] Beutner, A., Pettersson, S., Segl⍰, F., Nilsson, A., Bartels, A., Binh To, T. … & Hoess, A. (2011). *HAVEit deliverable 13.1 optimized and validated demonstration vehicles.* Retrieved from http://www.haveit-eu.org/LH2Uploads/ItemsContent/ 24/HAVEit_212154_D13.1_Public.pdf

[14] Beutner, A., Pettersson, S., Segl⍰, F., Nilsson, A., Bartels, A., Binh To, T. … & Hoess, A. (2011). *HAVEit deliverable 13.1 optimized and validated demonstration vehicles*). Retrieved from http://www.haveit-eu.org/LH2Uploads/ItemsContent/24/ HAVEit_212154_D13.1_Public.pdf

[15] Step by step towards success. (n.d.). *Smart Senior.* Retrieved from http://www1.smart-senior.de/enEN/Ergebnisse/Module

[16] The SARTRE project. (n.d.). *SARTRE.* Retrieved from www.sartre-project.eu

[17] Defense Advanced Research Projects Agency (DARPA) Urban Challenge. (n.d.). *DARPA.* Retrieved from http://archive.darpa.mil/grandchallenge/

[18] Nothdurft, T., Hecker, P., Ohl, S., Saust, F., Maurer, M., Reschka, A. & Bohmer, J. R.. (2011, October). Stadtpilot: First fully autonomous test drives in urban traffic. *Proceedings of the 14th International IEEE Conference on Intelligent Transportation Systems (ITSC)* (pp. 919 – 924). Washington, DC. doi: 10.1109/itsc.2011.6082883 October 2011.

[19] CyberCars – 2 (n.d.). *CyberCars2.* Retrieved from http://cybercars2.paris-rocquencourt.inria.fr/

[20] Audi of America (2013, January). *Piloted driving.* [Press release]. Retrieved from http://www.audiusanews.com/pressrelease/3304/149/piloted-driving

[21] Williams, M. (2012, October). Nissan previews self-driving car. *Computer World.* Retrieved from http://www.computerworld.com/s/article/9231978/ Nissan_previews_self_driving_car

[22] Thiesen, B. (2011, November). *Autonomous mobility appliqué system.* Presented at the SAE 2011 Intelligent Vehicle Systems Symposium – Advancing the Connected Mobility Experience, Troy Michigan. Retrieved from http://www.sae.org/events/training/symposia/ivs/presentations/2011/bernard_theisen.pdf

[23] Sheridan, T. B. & Verplank, W. L. (1978). *Human and computer control of undersea teleoperators* (Technical report). Arlington, VA: Office of Naval Research.

[24] Lee, J. D. & See, K. A. (2004). Trust in automation: Designing for appropriate reliance. *Human Factors: The Journal of the Human Factors and Ergonomics Society*, *46*(1), 50-80.

[25] Lee, S. E., Simons-Morton, B. G., Klauer, S. E., Ouimet, M. C. & Dingus, T. A. (2011). Naturalistic assessment of novice teenage crash experience. *Accident Analysis & Prevention*, *43*(4), 1472-1479.

[26] Rodriguez, S. (2012, November 30). L.A. Auto Show: Animal detecting systems coming as soon as 2013. Los Angeles Times. Retrieved from: http://articles.latimes.com/2012/nov/30/autos/la-fi-hy-autos-laas-animal-detection-night-vision-autoliv-20121129

[27] Audi TJA: Retrieved from: http://blog.caranddriver.com/we-go-for-a-ride-in-audis-piloted-driving-autonomous-a6-avant-prototype-2013-ces/
[28] Center for Internet and Society, Stanford University. Automated Driving: Legislative and Regulatory Action. Retrieved from: http://cyberlaw.stanford.edu/wiki/index.php/Automated _Driving:_Legislative_and_Regulatory_Action
[29] National Highway Traffic Safety Administration (2014). *New Manufacturers handbook: Requirements for manufacturers of Motor Vehicles and motor vehicle equipment*. Retrieved from http://icsw.nhtsa.gov/cars/rules/maninfo/newManf.pdf

SECTION 3. DEFINITION OF THE CURRENT SYSTEM

The baseline definition of the current system provides a reference point from which a discussion of change can occur, including the specifics of that change. This section provides that baseline definition for all stakeholders.

Operator-vehicle Interaction System Operational Definition

A ConOps document traditionally presents a framework documenting a system. The ConOps also provides a method to systematically define each important attribute of the system to an audience that might operate the system. For the present study, this ConOps document follows the framework of a traditional ConOps document but is intended to capture operations of a research/experimental system rather than a production system. For the purposes of scoping future research—and based on existing systems—this ConOps document provides an overview of the conceptual L2 and L3 systems with sufficient detail to avoid confusion while allowing enough variability that future technological advances may still apply.

This ConOps document covers automated vehicles that fall within NHTSA L2 and L3 definitions and is intended to provide an assessment of the details specific to these two levels of automated vehicles. For context, some L4 systems are briefly referenced as well. To ensure a common understanding of the automated vehicles specific to this research effort, there is a need to provide an operational definition of current motor vehicles and their automation technology. As subsequent research and development of L2 and L3 automation technology has the potential to dramatically change motor vehicles in combination with the legacy vehicles that will remain in-service, the definition should focus on the broad concepts of the operator-vehicle system. What is proposed is a general model and definitions that should apply to all motor vehicle operations in regards to the operator-vehicle interaction (OVI). This model and the definitions only examine the interaction level; they are not concerned with the operation of the dependent components. Instead, this model provides a succinct overview of the OVI system. It is important to note that there is a hierarchical taxonomy for this ConOps document in regards to system nomenclature. Within the system there can be multiple subsystems; each subsystem comprises components that are grouped together for categorical reasons. Within subsystems, there are components that individually exist to serve a specific purpose. Each process or function defined and discussed in this document represents a component which, when grouped by subsystem and collectively considered, represents the greater system. This structured hierarchy is necessary to ensure clear understanding of the relationships between the various parts (subsystems, components, etc.) of the system.

Operator-vehicle Interaction Subsystems

The OVI system consists of three subsystems: the operator, the vehicle, and the environment in which the OVI exists. Within these subsystems, the operator comprises three major components, and the vehicle comprises three major components. Beyond recognizing the environment as responsible for many possible inputs to the system, further breakdown of the environment subsystem is considered outside the bounds of the OVI system.

Within the operator subsystem there are Operator Input, Cognition Process, and Operator Output components. All external stimuli affecting the operator begin in the Operator Input component. Information is then passed to the Cognition Process, where the information processed by the operator and any needed responses are formulated. When a decision to act occurs, the directive for action is passed to the Operator Output component, and the action is implemented.

Within the vehicle subsystem, the components include Vehicle-Control Manipulation (VCM), Computational or Operational Process (COP), and Feedback or Display. The VCM component handles all inputs to the vehicle subsystem. Information is then passed to the COP component, where the input is processed for type, capability, and action. The actions are then passed to two locations: the Feedback or Display component, where acknowledgement of the subsystem input and subsequent action are communicated to the operator, and to other systems performing work. For the purposes of the operational definition of the OVI system, these systems (grouped here as non-OVI work systems) are considered external to the OVI. It is sufficient to understand that these actions are performed by the vehicle and can generate environmental feedback that re-enters the bounds of the OVI system through the environment subsystem. Figure 3-1 provides a model of this basic OVI system.

An example of this system is driving along a straight highway in clear daylight conditions. The operator receives environmental inputs in the form of visual information associated with the road conditions, lanes, traffic signs, other vehicles present, visual flow and direction of travel, and other factors. These inputs are continuously passed to the Cognition Process component for constant evaluation of the condition against the desired state of the vehicle. If the road begins to curve, this change is an environmental input to the operator subsystem. The Operator Input senses this curvature, the Cognition Process determines the need to adjust the vehicle heading, and the information is acted upon by the Operator Output. The Operator Output in this case becomes the hands on the steering wheel, changing the steering angle appropriately (determined through many iterations of this process during a short amount of time). The VCM – in this case, the steering wheel – provides a command input to the vehicle, thus prompting the COP[†] to appropriately act on the necessary non-OVI work system (e.g., steering wheel to pinion to rack and through tie-rods to steering knuckle). Some feedback is provided to the operator in one or more ways. In this example,

[†] Note: The Computational/Operational Process can be mechanical components, electronic components, or a combination of both.

the steering wheel angle change is a visual form of feedback in combination with the proprioceptive feedback of arm rotation and centrifugal force on the operator. Environmental feedback is provided through visual cues of vehicle heading change. These general processes continue through similar relationships during the driving experience.

Modes of Operation for the Current System

Current passenger vehicles vary in design, features, and age. Many of the features included in both the legacy vehicles and the latest models qualify for the L1 automated vehicle definition. Cruise control, for example, is a common feature allowing the human operator to cede control of the vehicle speed but requires the continuous operation of all other driving controls. Recent technologies such as lane keeping also qualify for the L1 definition. In these cases, as well as any other examples of L1 features, the operator is ultimately responsible for control of the vehicle, including maintaining situational awareness. The OVI system can function with these features either disengaged or engaged. The mode of operation while all features are disengaged is the normal operating mode of the system. However, within appropriate constraints, the system can be operated at the L1 mode by engaging one of these features.

Mode selection is primarily the responsibility of the operator within the system. The operator can choose to move the system from L0 to L1 by engaging an L1 feature, or the operator could move from L1 to L0 by disengaging the feature. Alternatively, the vehicle could detect a subsystem failure in the L1 feature and force the system to operate at L0. The current system is designed to function in either mode and requires the operator to make the appropriate decision as to which mode is best during system operation given environment, vehicle, and operator conditions.

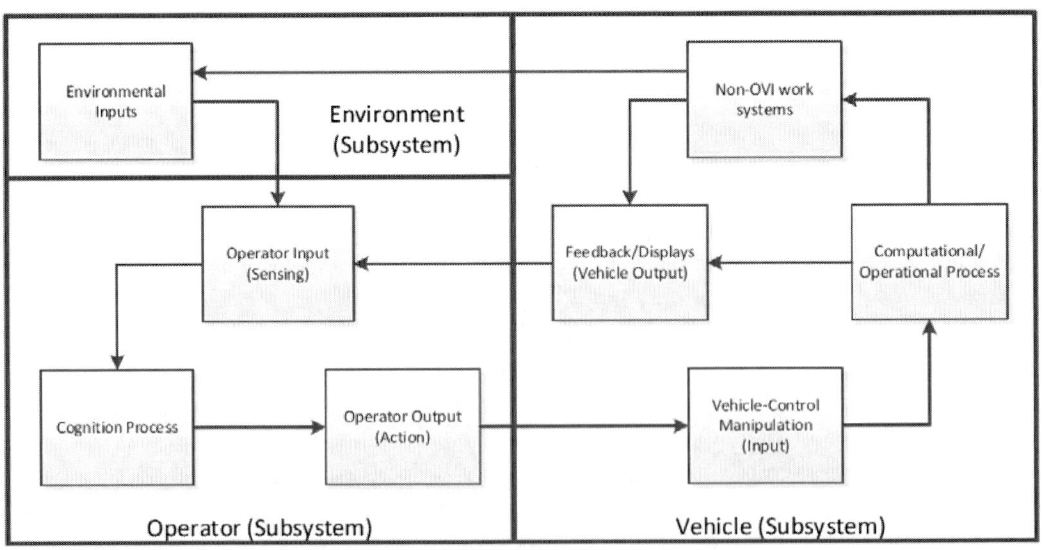

Figure 3-1. Operator-Vehicle Interaction System (Source: Author).

SECTION 4. LEVEL 2 AND LEVEL 3 SYSTEMS

As stated, the purpose of this ConOps document includes defining the levels of automated vehicles and framing the research problem objectives for L2 and L3 systems to inform subsequent research and development efforts. The operational definition herein provides a common language with which to structure discussion and, ultimately, possible future research.

In comparison to the current OVI system (L0 and L1), L2 and L3 automated components place a greater emphasis on the management of the system as opposed to the operation. That is, along the continuum of vehicle control and decision-making, L2 and L3 OVI systems place greater control and decision-making on the vehicle subsystems and automated components. The operator retains ultimate authority of the vehicle but delegates well-defined tasks to the automated components. The operator's interactions with the vehicle subsystem tend towards the management of these components through processing inputs from the environment and vehicle subsystems.

Level 2 and Level 3 Automated Systems Operational Definition

The overall set of automated vehicles includes a variety of features that can range from safety features requiring no driver intervention to vehicles that do not require an operator. As mentioned in the Introduction, NHTSA has developed a five-level taxonomy of automated vehicles, with specific performance-based definitions for each of the five levels. This ConOps is focused on L2 and L3 automated vehicles. It should be noted that these L2 and L3 systems are defined in a manner general enough to allow for future technological advancement while maintaining sufficient detail to provide clear boundaries for discussion, research, and development efforts on the same systems.

With respect to the definition of the OVI system, the one key addition for the automated vehicle, regardless of the level, is the Automation Component. The Automation Component is a general term that describes those technologies that provide some sort of automation to the vehicle operation system. Although the individual subsystems that compose automated vehicle controllers are highly complex, they are not the focus of this model and can be represented in terms of a single addition to the model. The operator can choose to engage the Automation Component or operate the vehicle in a traditional (i.e., manual) manner. The Automation Component can act as a supervised agent of the operator in that environmental inputs and vehicle outputs can be acted upon by the Automation Component to continue operating the vehicle in a manner consistent with the operator's intent.

Thus, in an L2 system, the Automation Component can either supplement the operator or relieve the operator of one or more subtasks of the overall driving task. Examples of L2 vehicles are those that combine lane-keeping and adaptive cruise control (ACC) operations during which primary lateral and longitudinal control is ceded by the operator yet direct supervision is still required from the operator. The operator continues to receive system feedback and is responsible for much of the overall operation of the vehicle. In L3 vehicles, the Automation Component acts in a manner that removes the direct supervisory requirement and changes the operator's role to being available for control with some degree of notice.

Figure 4-1 provides a representation of the OVI system with Automation Components for L2 and L3 automated vehicles.

One can consider lane keeping as an example Automation Component. When the driver engages the lane-keeping feature of the system, the system then acts on behalf of the driver to maintain the vehicle position within the travel lane. In this example, certain information about the vehicle is continuously provided to the operator, including confirmation of lane-keeping activation. However, assuming all else remains relatively static, the Automation Component acts to make corrections via the Vehicle-Control Manipulation component based on environmental inputs (e.g., machine vision verification of relative position of vehicle to lane lines). Each correction is fed through the COP and, as a result, is provided back to the Automation Component. This cycle continues until the operator resumes manual control or the Automation Component determines that it is outside the bounds of the system design. Throughout this cycle, pertinent information is provided to the operator through the Automation Component to the Feedback or Display component, thus allowing the operator to maintain an active role in the OVI system.

It is important to note that environmental inputs act on both the Automation Component and the Operator Input components. However, the operator still holds ultimate authority over, and control of, the system.

The operator has at all times the option to act through the VCM component. This action by the operator can bypass the Automation Component or could have a dual action (e.g., depressing the brake may turn off adaptive cruise control while simultaneously acting to slow the vehicle).

Lastly, the system as defined above is for a conceptual idealized process. In all systems there is the potential for component failure. The frequency and effects of component failure are variable and will not be defined for this proposed OVI system. However, it is important to understand that this proposed OVI system is no different and has the potential for failures, specifically in the feedback loops. For instance, should the Automation Component fail to act in a manner consistent with the perception of the operator, there may be no notification (for L2 systems) or awareness (for L2 and L3 systems) of this failure, which could result in an accident. While this idealized OVI system definition provides a baseline for discussion, it is expected that subsequent research will focus on these potential issues to characterize, understand, and propose solutions for the production of L2 and L3 systems.

Exposition of System Prototypes under Active Development

The purpose of this section is to gain a concrete sense of what systems are more likely to be introduced to the public. The industry has been very active in recent years in demonstrating prototypes of various automated vehicle concepts and in bringing early systems to the market. This section describes system approaches that have already moved into production or have the potential to do so during the coming years.

This section is not intended to be exhaustive; rather, it is intended to be indicative. It includes only automated vehicle research work in which the vehicle industry was involved.

This information will feed into later sections so that the selected concepts are relevant to likely future market offerings.

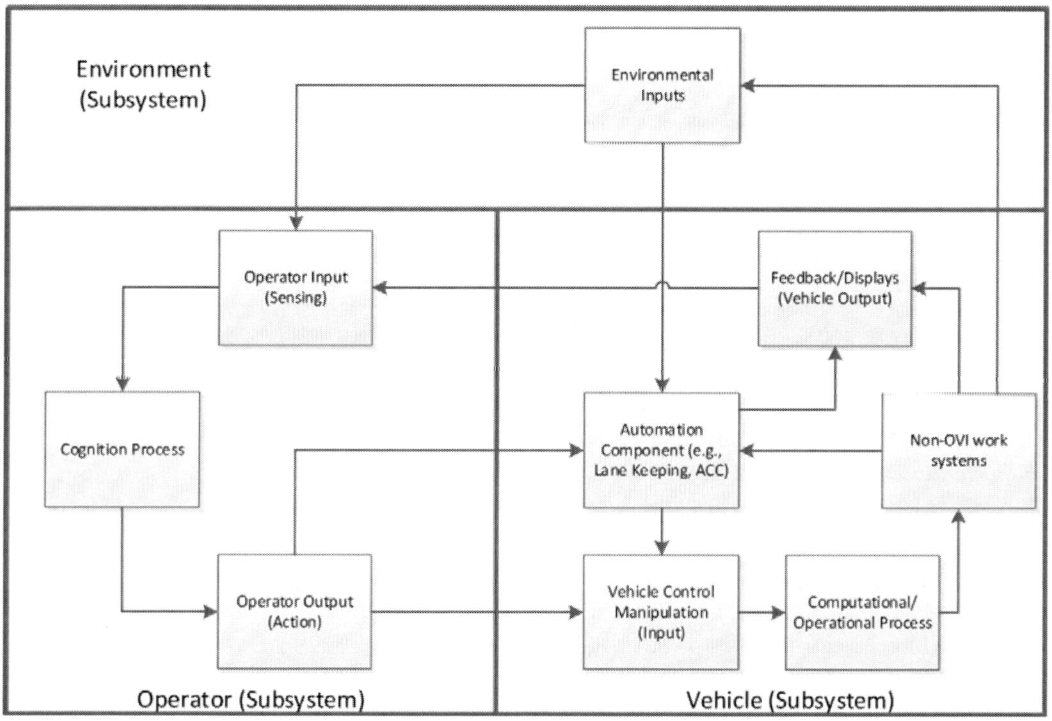

Figure 4-1. Automated Operator-Vehicle Interaction System (Source: Author).

Traffic Jam Assist

For model year 2014 (MY14), both Mercedes and BMW have announced the availability of Traffic Jam Assist (TJA). TJA functions on limited-access highways at slow speeds. Building on Full Speed Range Adaptive Cruise Control, the system provides full control of driving in congested conditions. Press announcements for the availability of TJA by "mid-decade" have also been made by Audi, Ford, and Volvo.

The Mercedes S-Class system is representative of TJA.[7] The sensor suite is enhanced with stereo vision, which enables normal lane marker and car-following detection when in a traffic jam. Speed is adapted to the flow of traffic ahead, slowing as necessary. The system can brake to a full stop. When traffic moves, the driver can resume with a tap, or, if the stop is less than 1 s, automatically.[8] The driver is expected to be engaged, with his or her hands on the steering wheel. If the driver is not touching the steering wheel, the system issues a warning and will disable after a few seconds.

Within the European HAVEit project, the AQuA demonstrator was developed on a heavy-truck platform.[8] The AQuA function supported a truck driver in congested traffic by autonomously handling the speed and steering control. The AQuA aimed to relieve the driver of the monotonous tasks associated with driving a truck in congested traffic situations; that is, driver underload situations. The level of automated control was continuously adapted across L0 to L2 systems based on the states of the driver, the vehicle, and the environment. At the highest level of automation, the system autonomously handled steering, acceleration, and braking to keep the vehicle in the correct lateral position in the lane and at a safe distance from the preceding vehicle or at a desired speed. The system operated at speeds between 0 and 130 km/h.

Highway-Speed Automation

The General Motors (GM) "Super Cruise" system[10] is described in press releases as providing full-speed-range ACC and lane keeping, using cameras and radar to automatically steer, accelerate, and brake in highway driving. The GM system allows the driver to take his or her hands off the steering wheel until the driver wishes to change lanes, road conditions deteriorate beyond system functionality, or other system issues occur.

Infiniti has announced that MY14 vehicles will be equipped with "steer by wire," which allows independent control of the tire angle and steering inputs.[11] This capability insulates the vehicle from unnecessary road-generated disturbances such as ruts or cross winds. This system enables a new "straight-line stability system" that assists "on-center driving." The system reduces discrepancies between the intended path and the actual path. In addition, the company asserts that fewer fine-grained steering adjustments made by drivers reduce their fatigue/workload.

In Europe, the Honda Accord is available with a Lane Keeping Assist System (LKAS) that provides hands-off lane keeping for a limited period of time (tens of seconds). The vehicle can also be equipped with ACC, making simultaneous LKAS/ACC possible.

Audi offers an LKAS with an "early intervention" setting, which in effect provides a lane-keeping function that can be operated simultaneously with ACC. It requires the driver's hands be on the wheel and disables if they are not. This version of the system is only available in Europe.

BMW[12] has demonstrated an L2 vehicle that provides lateral and longitudinal control at highway speeds. When slower traffic ahead prevents the vehicle from maintaining its set speed, the vehicle monitors the adjacent lane and performs a lane change when safe. The vehicle also responds to merging traffic on the right by adjusting speed.

Within the European HAVEit project, the Joint System Demonstrator (JSD)[13] was developed that provided L0 through L2 system automation. The vehicle was equipped with warning-/intervention-based active safety, ACC, and LKAS functions. Steering wheel buttons allowed drivers to choose a specific automation level. Detection of the driver's hands on the steering wheel was implemented.

Google has developed an automated vehicle that currently can operate up to 75 mph at L2 in highway use and L3 for safe testing purposes (e.g., research). The vehicle does not change lanes. It will slow down or stop for moving traffic in front of it. The vehicle will stay in its lane in automated driving mode until the user is ready to retake control through a combination of virtual mapping, laser, radar, and cameras. An in-dash user interface displays system state information and is used to notify the driver when automated mode is available, as well as to prompt the driver to retake control when needed.

Automated Assistance in Roadwork and Congestion[14]

The Automated Assistance in Roadwork and Congestion (ARC) demonstrator was developed within the European HAVEit project. The main focus was driving through a work zone in the highly automated mode (L2). This scenario could be an overload situation at speed or an underload situation if traffic is congested.

The roadwork assistant application considered the possibility that lane lines may not be visible or accurate. Therefore, additional objects such as trucks, beacons, and guide walls were used for guidance. A control algorithm called "virtual wall" steered the vehicle back into the lane when it got too close to a conventional or unconventional lane border. Otherwise, an LKAS provided lane keeping, and ACC provided longitudinal control. The automation operated at speeds between 0 and 80 km/h.

Emergency Stopping Assistant[15]

BMW developed this system for situations during which a driver becomes completely impaired at the wheel (e.g., a heart attack or stroke). An L4 system, the Emergency Stop Assistant, detects the situation of an incapacitated driver, turns on the vehicle hazard signals, and safely maneuvers the vehicle through traffic to park on the roadside.

On-Highway Platooning

Within the European SARTRE[16] project, Volvo cars and trucks were equipped for close-headway platooning (L3). The project investigated the business model of a human driver being paid to drive the lead vehicle by the occupants in the fully automated following vehicles in the platoon. Prototypes were developed using current production car/truck technology plus Dedicated Short-Range Communication (DSRC) for V2V communications. The cars drove automatically behind the lead vehicle at 85 km/h, separated by 5 to 15 meters. Several hundred kilometers of testing were performed.

City Street Automated Driving

The 2007 Defense Advanced Research Projects Agency (DARPA) Urban Challenge attracted dozens of teams with vehicles which could operate among each other in a residential street setting. The top three finishers (Carnegie Mellon University/General Motors, Stanford, and Virginia Tech, respectively) successfully handled traffic circles, four-way stops, parking lots, and road obstacles on designated streets.[17]

The University of Braunschweig in Germany built upon its entry into the DARPA Urban Challenge with its Leonie vehicle (L3).[18] This research focuses on urban driving, and the university regularly conducts testing in live traffic in the city "Stadtring," a busy four- or more-lane arterial with traffic lights, turn lanes, pedestrians, etc. The vehicle executes turn maneuvers, responds to traffic lights, and performs lane changes as needed to flow within traffic. Partial funding for this work comes from the vehicle industry. The European CyberCars[19] project has developed urban vehicles for L4 operation within city centers in lanes or districts segregated from normal traffic. Long-term demonstrations have been held in several European cities.

Automated Valet Parking

Audi[20] has developed a prototype system that enables the driver to depart the vehicle at a parking garage entrance and to instruct the vehicle to "go park" via a smartphone interface. The vehicle then proceeds, empty, into the garage, travels until it detects an empty space, and parks. Upon being summoned by the owner via smartphone, the vehicle proceeds to the garage exit area where the driver re-enters the vehicle and resumes driving. Nissan[21] has demonstrated a similar system.

Table 4-1. Summary of Automated Vehicle System Products/Prototypes under Development by Industry

Operating Environment	System Function	Level	Status	0–5 years	5–10 years	10–15years
Highway	Traffic Jam Assist	2	Product announced for MY14	√		
Highway	On-Center Driving w/ACC (limited steering authority)	2	Product	√		
Highway	Combined lateral/longitudinal control	2	Prototype	√		
Highway	On-Highway Platooning	3	Prototype		√	
Highway	Workzone Assistant	2	Prototype		Note 1*	
Highway	Emergency Stopping Assistant	4	Prototype		√	
City	Segregated Urban Automated Vehicles	4	Trial Deployment	√		
City	Open Road Urban Automated Vehicles	3	Prototype			√
All Roads	Automated Operation on All Road Types for Military Applications	3	Prototype		√	
Parking	Automated Valet Parking	4	Prototype	√		

* Note 1: Workzone Assistant is expected to be integrated into full highway functionality when such products emerge.

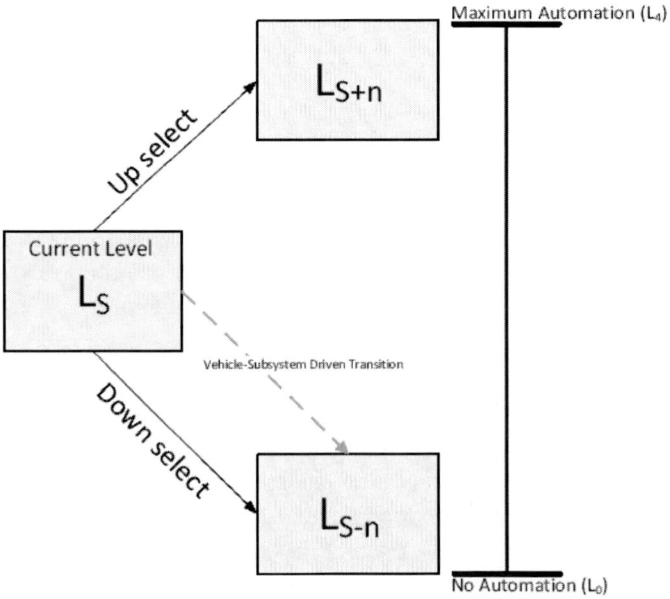

Figure 5-1. Transition of Automated Levels Model (Source: Author).

All-Road Autonomous Operation

The U.S. Army[22] is sponsoring development of the Autonomous Mobility Appliqué System, a program designed to retrofit existing military trucks with a range of systems, from active safety to full automation (L3). The intent is for the vehicles to operate on any road type and off-road. BAE Systems, which produces military trucks, is the vehicle industry participant.

SECTION 5. MODES OF OPERATION

Model of Transitions

Many systems are designed to operate across multiple modes. There are a number of reasons for including this capability, including maintenance, failure, and process agility. The OVI system as defined in this document is multimodal. For the specific system considered herein, the two key modes of interest are when the vehicle subsystem is operating at (or transitioning into/out of) L2 or L3 automated driving. While alternative modes of operation are outside the bounds of the system in consideration for subsequent research, the transition between these two modes is of interest. Additionally, the system can transition from one level to another. Such transitions may include increasing from L0 or L1, to L2 or L3, or transitioning between levels. A transition from one level to another could be initiated by either the operator subsystem or the vehicle subsystem based on inputs to either. Figure 5-1 provides a static look at the dynamic system state of a multi-level autonomous vehicle. The current state of the system is always represented as L_S. The transition to a higher level of autonomy (up-select) is represented as L_{S+n} with a maximum of L4. Transitioning to a lower level of autonomy is represented as L_{S-n} with a minimum state of L0. A transition can be made from any level of autonomy to any other level (e.g., L0 to L3 or L2 to L1) depending on the design and function of the system. System transition can be initiated through an action of the operator subsystem or vehicle subsystem for any number of reasons.

The transition between levels of automation can be complex. For instance, the operator starting a vehicle and manually shifting into gear to move the vehicle forward is an example of L0 automation. The driver has authority and responsibility for all movement and control of the system and components. If the driver chooses to engage ACC upon entering a highway lane, the overall system transitions to L1 automation. The driver (operator subsystem) has ceded limited control and authority of speed maintenance to a vehicle component (i.e., vehicle subsystem automation component). Transitioning through the model, assume the driver then engages a secondary automated subcomponent (e.g., lane centering) while still using ACC. Now the OVI system is operating at L2 automation. Conceivably, this could continue through full automation if the vehicle were so equipped. Although these up-transitions are rather straightforward, the down-transition process can be more complex from a human-automation interaction perspective.

For example, assume a system operating at L2 exceeds an operating parameter within either the lateral control or speed regulation component. Assuming both engaged automated components are capable of recognizing that the environment is no longer sufficient for supporting safe operation at L2, the vehicle subsystem will cede control to the driver (discussion about situations where the driver does not accept control can be found in the *Misuse in L2 and L3 Vehicles* and *Terminal State* sections below). When this transition occurs, the level transition is from L2 to L1 or L0, depending upon whether one or both of the components are disengaged. Therefore, OVI systems are, and will continue to be, capable of transitioning both one level at a time as well as multiple levels at once. This holds implications for how the warnings and alerts are developed and may have implications as to whether the operator understands the transitions completely before or as they occur.

Shared Authority

A key area of interest for future research is the process of transitioning between the various levels of automation and the concept of shared authority (or control).[23] Shared authority refers to the breakdown of all tasks necessary to operate the vehicle (i.e., complete task list for the OVI system) assigned to both the operator subsystem and the vehicle subsystem. While L2 operations are designed for the human to be present, to monitor the automation performance, and to be capable of taking control at any point, L3 operations are designed for the human to be present and available for occasional control given sufficient notification. Therefore, each level of automation has the capability to transition to an L0 vehicle that places all authority on the operator. However, some functionality of the vehicle will, at times, operate as shared authority.

Proper functioning of the OVI system in L2 and L3 systems requires some understanding of task assignment. For L2 systems, the operator subsystem is expected to monitor the roadway and be capable of taking over the entire set of driving tasks at any time. In L3 systems, shared authority is more coarse, requiring only that the operator subsystem understand when the vehicle subsystem is operating, and when provided sufficient notice take control for safe vehicle operation back from the vehicle-subsystem.

Communication and Feedback Loops

The OVI system can be considered somewhat decentralized based on the three major subsystems that compose the greater system. The two subsystems with control capabilities are the operator subsystem and the vehicle subsystem. While each of the three subsystems is integral to overall system functionality, there is a difference in communication across each of the systems. The environment subsystem has one-way information input (please note that this does not negate the fact that changes in the vehicle subsystem can affect the input from the environment to the other subsystems). Thus, no information from the other two subsystems will affect the environment subsystem. It simply exists and, as such, must be accounted for in the performance of the other two subsystems. Conversely, the operator and vehicle subsystems communicate with scenario-dependent variable frequency. This works in both directions, and failure of one of these communications segments presents the opportunity for OVI system conflict (e.g., failure to engage an automated component or physical conflict with an environmental input). The vehicle subsystem provides information to the operator, who may take actions that become feedback mechanisms. The simple model in Figure 5-2 summarizes the communication flow between subsystems.

Within the OVI system, the communication between subsystems should be unambiguous. The driver must take a specific action in order to result in a given vehicle output.

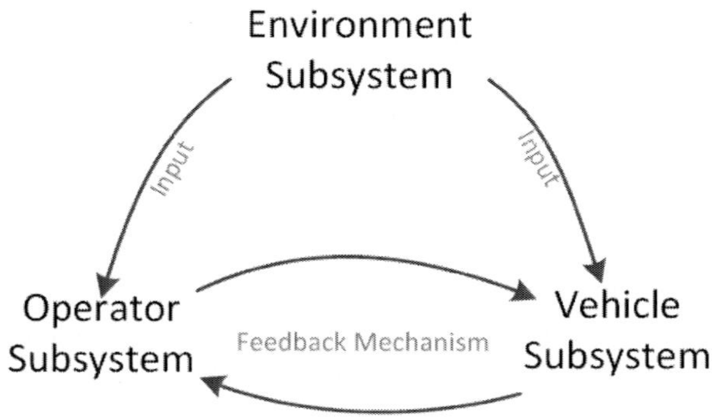

Figure 5-2. OVI Subsystem Communication Model (Source: Author).

Engaging the Automated Component(s)

Performance

When the operator chooses to engage an automated component, the operator has certain expectations. The operator may expect the system to engage immediately, or the operator may expect the automated component to recognize an appropriate operating environment prior to fully engaging. The concept of performance is twofold. It refers to the operating performance of the automated component of the vehicle subsystem, and it refers to the operator's interaction with the automated component.

Operation of an L2/L3 vehicle, or any vehicle for that matter, requires that a certain number of tasks be accomplished. A complete task list for the operator of a motor vehicle is long and complex. However, there is a base set of tasks necessary to safely and effectively accomplish the objective of operating the motor vehicle. In an L0 vehicle, each of these tasks is the responsibility of the operator. In a hypothetical L4 automated vehicle, each of these tasks would be the responsibility of the vehicle (conceivably, the operator is always capable of input but, by definition, this would not be necessary). Therefore, L2/L3 vehicles will exist somewhere between the two extremes of task responsibility. There will be a shared task authority, which will be dependent upon the level of automation and the choices of the operator to engage or disengage the automated component(s).

For safe and effective operation of the OVI system, it may be beneficial if the operator understands key tasks for which he/she has authority. There is a temporal aspect to this issue as well. The operator can choose to engage (or disengage) the automated component at any time, assuming a supportive environment (i.e., an environment within the requirements for the automated component to function properly as defined by the manufacturer). Thus, operator comprehension of the OVI system includes which tasks are assigned to the operator and when those tasks are the responsibility of the operator. Subsequent research about the operator's comprehension of task authority (specifically when transitioning from one level of automation to another) could be of benefit to future OVI system design.

To illustrate this concept, consider the use of two automated components: lane keeping and ACC. When the operator engages the cruise control, he/she is ceding task authority for maintaining speed. The operator has the ultimate authority in that he/she determines the set speed, engages the automated component, and can disengage when he/she chooses. However, while cruise control is engaged, there is no need for the operator to make constant subtle adjustments to the throttle (accelerator pedal) based on environmental factors (e.g., external inputs such as terrain, wind, traffic, etc.). Should the operator also engage a lane-centering automated component, he/she is also ceding task authority for maintaining vehicle heading. The constant adjustments for the vehicle position in the lane are left to the automated component within the vehicle subsystem. Therefore, two high-level task categories are controlled by the vehicle subsystem.

Consider an additional example: A driver engages a lane-keeping automated component. Subsequently, the driver chooses to change lanes. After manually changing lanes, the driver assumes that the lateral control functions of the lane-keeping automated component will re-engage. The system is not designed to re-engage without driver input, thus leaving the driver responsible for lateral control. This example of potential mode confusion could have the potential to lead to a safety-critical event. Research to understand the operator subsystem expectations and interactions with system design may provide greater insight into the possibility of mode confusion when engaging and disengaging automated components.

Trust in L2 and L3 Automated Vehicles

Automation across any domain has the potential to present difficulties when users' reliance upon the automation is not properly calibrated to the performance. The relationship between trust and automation can lead to several different outcomes, as defined by Lee and See.[24] Calibrated trust describes a system in which the user's trust matches the automation capabilities. Calibrated trust supports appropriate application of the automation. Over trust describes a system in which the user's trust in the automation exceeds the actual capabilities. Over trust can lead to misuse of the automated system, where the driver applies the automation to a roadway environment that is outside the automation operational scenarios. Distrust describes a scenario in which the user believes that the automation performance is less than it actually is. Distrust can lead to disuse of the automation, thus removing the possible benefits of the automation.

Although trust in automation has been studied across a number of domains, the applicability of these findings to the driving task is questionable due to fundamental differences between the operational characteristics of the different automated systems. Perhaps the best analogy may be drawn from aviation. Modern aircraft have multiple highly automated systems onboard. Modern autopilot systems have the ability to automate almost all phases of flight. This allows the pilot to transfer to a monitoring role. However, the analogy begins to fail when considering the fact that airspace is somewhat controlled, routes are mapped, and the potential for other airspace users to perform unexpected threatening maneuvers in very close ranges (i.e., less than 2 seconds) is not typically present. Therefore, while the lessons learned from other automation domains should be considered when approaching L2/L3 vehicle automation, they may not necessarily be directly applicable. More research about drivers' trust in L2/L3 automation can be beneficial to OVI system designers and stakeholders. Ultimately, the issue of drivers' trust in L2 and L3 automated vehicles is critical to how these technologies are both used and misused in the real world.

Misuse in L2 and L3 Vehicles

The concept of misuse within the context of automated vehicles and the aforementioned OVI system can be defined as the operator subsystem, when having knowledge and understanding of the automated component alert, cue, or other interaction, deliberately chooses not to act according to the intent and design of the automated component. That is, should a driver fully comprehend the alert provided by the L2 system and choose to ignore it for his/her reasons, it is a case of misuse. Cases when the operator subsystem does not comprehend or receive the alert (feedback from the vehicle subsystem) or confuses the alert for something else, are not misuse. The concept of misuse is independent of safety or convenience and hinges upon the knowledgeable choice of the operator based on his/her goals at the time of the decision.

In L2 vehicle systems, the operator subsystem, by definition, is expected to monitor the roadway and safe operation of the vehicle while remaining available to take full control at any time. Therefore, when the operator subsystem, potentially having full knowledge of this responsibility, chooses to engage in a secondary task that may disrupt or eliminate operator capability to effectively monitor the roadway and safe operation of the vehicle, this can qualify as or lead to misuse. Conversely, in an L3 system in which, by definition, the operator subsystem has sufficient time to transition from a separate task back to the driving task, the purposeful engagement of an alternative task is not necessarily misuse. However, in an L3 system, misuse could occur when the operator subsystem attempts to ignore warnings/alerts fully comprehended (including potential consequences for ignoring system communication) by the operator for reasons known to the operator. Misuse can occur in many different ways but always requires the operator to intentionally act in a manner inconsistent with, and with full knowledge of, the design of the system.

Disengaging the Automated Component(s)

Both the operator subsystem and the vehicle subsystem are capable of disengaging the automated component from operation. There is an unequal interaction in that the operator may typically expect immediate response to disengaging an automated component, whereas the vehicle subsystem will often provide a warning or alert process to attempt to facilitate operator awareness in preparation for disengaging. Each case is explored independently below.

Operator Chooses to Disengage

From an operator's perspective, the use of an automated component provides relief of some responsibility for driving tasks, the extent of which is dependent upon the type and number of automated components available within the OVI system. Evaluation, exploration, and understanding of operator disengagement are considered within three categories: Performance, Comprehension, and Misuse.

Performance

Performance refers to how well the interaction between subsystems functions when the operator chooses to disengage the automated component(s). The methods by which the operator provides disengagement input and the subsequent feedback loop can be measured to quantify performance. Specific to automated vehicles, performance is defined as how well the operator subsystem interacts with the vehicle subsystem. In cases of L2/L3 vehicles, performance considers the timing, reaction speed, and clarity of understanding between the operator and vehicle in modes of transition from one level of automation to another (or even changes within a level of automation).

Comprehension of Authority

A key issue with disengagement is whether the operator sufficiently understands the additional task authority placed on the operator subsystem. For example, assuming the OVI system is already functioning as an L2 automated vehicle, consider the act of disengaging ACC and lane centering. The operator must first provide a disengagement input (e.g., button press on dashboard) informing the vehicle subsystem that the operator is willing and prepared to take over responsibility for maintaining speed and/or heading. The vehicle subsystem may then acknowledge the input with a feedback mechanism(s) that provides confirmation to the operator. A time period then passes before the automated component disengages, and the operator becomes responsible for speed and heading.

There may be potential for operational issues if the operator is not aware of certain specific tasks for which he/she is accepting responsibility. Should the operator believe he/she has only accepted responsibility for speed maintenance, for example, a potential conflict arises regarding authority for vehicle heading. Essentially, the operator may incorrectly assume the vehicle is still maintaining heading when, in fact, neither subsystem is actively steering the vehicle. This concept of mode confusion is applicable beyond this simple example to other scenarios during which multiple automated components are engaged. Research to evaluate the operator's comprehension related to component disengagement can benefit system design, as can research to determine methods that ensure accurate feedback mechanisms and necessary information are provided to the operator.

Disengage Misuse

There is the potential for an operator to choose to disengage an automated component with the intent of operating the OVI system in a manner inconsistent with design. Although the reasons for this may be many and varied, there is value in identifying when misuse is occurring. There is a clear distinction between not comprehending the automated component functionality and intentionally misusing the system. As previously noted, discussion about misuse in this document refers to willful actions to use the automated component incorrectly. When the operator chooses to disengage a working automated component during a time or in a manner that the operator knows to be against the design and intent of the automated component, this is considered misuse. Subsequent research could focus on understanding misuse, characterizing misuse, and potential solutions for misuse. Dealing with misuse falls within Misuse Handling and is discussed in detail below.

Vehicle Disengages Automated Component

There are scenarios where the vehicle subsystem automated component can determine it is no longer capable of operating at a sufficient level of performance. In these cases, the automated component within the system may attempt to disengage and shift control authority for the task to the operator subsystem. This presents a unique case in the OVI system during which the communication originates with the vehicle subsystem. This case also serves as an example of when the operator subsystem should provide feedback to the vehicle subsystem to complete the communication loop. However, it is conceivable that the automated component can fail and disengage or could function properly but determine the need for an immediate disengagement. Subsequent research and further exploration of these scenarios could provide a benefit to system design.

Notification Cues

Notification cues herein are considered those feedback mechanisms provided from the vehicle subsystem to the operator subsystem when indicating something specific to the driving tasks. These may take the form of alerts/warnings to the operator when the vehicle subsystem intends to transfer some or all control back to the operator. While the types and specifics of the cues may vary between levels of automation and manufacturers, they are designed to indicate some type of message for operator assessment.

Subsequent research into the types and methods of notification generated through vehicle subsystem cues can inform future system design. There may also be some benefit to future system design from research into methods or techniques to handle ignored cues or to determine why the cues were ignored (e.g., potential misuse). When a cue is present but no operator subsystem action is taken, it may be of value to investigate the best course of action for the vehicle subsystem (e.g., increase in cue volume or secondary alert [e.g., haptic] supplemental to original alert [e.g., auditory]). In addition, research could provide input into the timing, sequence, and presentation of such cues for optimal system function.

Terminal State

The terminal state refers to the system actions when a dynamic loop reaches a static state. This concept considers the steps taken by the automation component when the operator subsystem fails to maintain or provide the feedback mechanisms within the overall system. If the automated component requests operator control and – for any number of reasons – the operator fails to regain control, the terminal state may need to be pursued. An example of this in an L3 vehicle would be when the automated component(s) detects a need for the operator to assume full control of the vehicle and sends an alert. The operator, for whatever reason, fails to disengage the automated component. While continuing to attempt to get the operator to comply, the system may have to bring the vehicle to a safe state. In both L2 and L3 vehicles, existing prototypes and designs indicate that many methods for handling the terminal state are in development. An example of placing the vehicle in a terminal state is having the automated component move the vehicle to the side of the road and come to a complete stop. This document and subsequent research could help inform this issue of safely placing the vehicle in a terminal state or, where applicable, preventing the need for such actions. This may include research by academics, original equipment manufacturers (OEMs), and others into developing and understanding fail-safe and fail-operational requirements for automated vehicles.

Misuse Handling

As mentioned, there is potential to misuse an L2/L3 system, specifically the automated component. Although misuse is intentional, it is likely the operator believes that misuse will not increase the risk of operating the vehicle. Regardless of his/her perception of this matter, the handling of misuse could have the goal of combating any potential increased risk. For example, if a driver chooses to engage a lane-centering system but fails to disengage the system when appropriate, the vehicle may detect the need to move steering authority back to the operator. During this scenario, the vehicle subsystem may attempt to notify the driver, although it is conceivable that in some cases it would not make such an attempt. If the driver is aware of and understands the request but does not take the appropriate action—believing instead that the vehicle will continue to operate in the same manner—this qualifies as misuse. This adds a third aspect to shared authority in that neither control subsystem is actively taking authority of a necessary task for driving (e.g., steering).

A key difference between the two control subsystems (i.e., operator and vehicle with automated component) is that the vehicle subsystem does not choose its task assignment. The vehicle subsystem handles tasks as assigned or may provide a cue (dependent upon system implementation and use-case scenarios imagined for system design) to the operator that the task can no longer be handled by the vehicle subsystem. However, the operator subsystem has the ability to understand key aspects of the tasks assigned and to choose not to handle one or more of them. The vehicle subsystem with automated components may then potentially have methods for handling this or other types of misuse. Evaluation and research have the potential to identify cases of misuse, to characterize these cases, and to develop functional countermeasures that may prevent misuse from increasing a safety risk to the OVI system, other vehicles, or infrastructure.

SECTION 6. RISK ASSESSMENT

As stated, this document covers the concepts of automated vehicles as they are developing and may be developed in the future. Some discussion is warranted at this stage in the development process about the issues associated with shared authority systems. It is important to note that these issues may be well understood and handled effectively by manufacturers and system designers; this will become clearer as development and production continue.

Potential Issues with Automated Vehicles

Potential issues with automated vehicles include, but are not limited to, the following.

Mode Confusion. As stated, the concept of shared authority in an automated vehicle considers the tasks assigned to the operator and to the vehicle. Within the OVI system established in Section 3 Definition of the Current System of this document, this concept is defined by the two control subsystems. An operational risk exists when both systems simultaneously perceive the other to be the ultimate controlling authority. This issue is discussed to some extent in Section 7 Modes of Operation under Comprehension of Authority. It is reiterated here to emphasize the potential benefits from a human factors perspective of understanding this risk and designing systems accordingly. Across all levels of automation, there is the potential for the operator to misunderstand his/her authority and responsibilities within the OVI system, which may be associated with a safety-critical incident. Research designed to qualitatively and quantitatively assess this risk as it pertains to automated vehicles has a potential benefit to system designers. Although a singular concept, there may be a variety of ways in which this risk could manifest itself within different system designs.

Error Handling. A second aspect associated with automated vehicles is defining intent and ceding ultimate control. It is conceivable that automated components within the vehicle subsystem will be designed to account for a certain amount of human variability. That is, people will make mistakes, and many of these mistakes will likely be understood. The designs of these automated components will account for such mistakes and will act accordingly. The risk in this approach is that the automated component may not correctly account for the intentions of the operator subsystem and may act inconsistently with the expectations of the operator subsystem. For a detailed explanation and discussion of this risk, the reader is referred to the philosophical differences of Boeing and Airbus approaches to automation in the Literature Review document. To summarize, the designers of automated components can provide ultimate decision authority to the vehicle subsystem, or they can provide ultimate authority to the operator subsystem. In the case of an automated component inadequately adjusting to an unusual event, the operator can detect the situation and control the vehicle accordingly.

To some extent, the issues here are similar to error handling in current crash-avoidance products (e.g., collision imminent braking). Further research into shared authority and transitions between controlling subsystems could benefit system design and could lead to a better understanding of the nuances of design philosophies in the realm of automated motor vehicles.

Overreliance. There is a possibility of overreliance upon automation when drivers are able to use an L2 or L3 automated driving system for some period of time. That is, the potential exists for drivers to constantly seek to activate the automated driving component of the vehicle after becoming familiar with the automation and how it functions. Researchers have postulated that, over time, this overreliance can possibly lead to degradation in driver skill as the reinforcement from constant engagement in the driving task is now lacking. This issue is of concern in the aviation domain, where pilot skill has been observed to suffer in the presence of frequent and continuous use of automated functions.

Driver Underload. Automation removes some demand from the driver and places it upon the automated system. However, by removing some level of task demand from the driver, there may be an associated risk of drivers achieving a state of underload. This is a situation in which the driver is under-stimulated, which may lead to fatigue, boredom, reduced levels of operator alertness, and sensation-seeking behaviors. This is potentially of more concern in L2 vehicles where there is an expectation that the driver will monitor the performance of the automation. In L3 vehicles, the expectation is that drivers may become more disengaged from the driving task. Longitudinal human factors research could provide insight into the issue of driver underload during extended periods of time when the operator becomes comfortable with the overall system.

Potential Benefits of Automated Vehicles

Reduced Workload and Fatigue. There are expectations that automation can reduce or eliminate certain risks to the current OVI system. Human factors research has long shown certain deficiencies in human capabilities to interact with machines. Specifically, humans have issues with vigilance; humans can suffer from decreased attention and visual acuity; and, under duress, humans often fail to account for all available cues. Automated components within the vehicle subsystem have the potential to augment the operator's capabilities in several ways. First, automated components can aid in vigilance. For example, automated components can remain a constant sensor for forward collision or ensure vehicle heading while allowing the operator a break. Within driving tasks, there is a finite workload to which the human can be subjected without suffering fatigue. While the number of inputs, tasks, and responsibilities of the operator subsystem are not entirely in control of the operator, automated components can decrease the total operator workload. This can, in turn, decrease fatigue, add convenience, and improve the overall OVI system experience from the perspective of the operator. Subsequent research has the potential to provide data about driver fatigue relative to operating an automated vehicle.

Increased Performance in Handling Small Errors. Both L2 and L3 operations present the potential for greater performance in handling small driving errors that humans typically commit. These errors are committed for a number of reasons (e.g., distraction, confusion, disengagement, or fatigue) and are generally below the level of safety concern. An example of a small error is a steering reversal that does not result in a lane exit. While the human may be subject to many factors that affect performance, it is assumed that any L2 or L3 production automated system will be able to provide a constant level of environmental monitoring and will be able to provide an appropriate response to the extent that the system has an accurate recognition of the driving scenario.

Driver Distraction. As mentioned, driver distraction can lead to small driving errors typical in L0 and L1 vehicles. However, driver distraction is also responsible for significant driving errors, including events leading to a crash. A large collection of research indicates that driver distraction is responsible for many crashes, injuries, and even fatalities. While L2 systems, by definition, require monitoring by the driver and offer more of a respite of the hands and feet in the physical act of driving, the L3 system may offer some protection from the issue of driver distraction. In an L3 system, by definition the driver cannot be distracted (from the driving tasks) when the automated component is engaged because he or she is not expected to monitor the roadway or be responsible for safety systems. Thus, activities considered secondary in an L0/L1 system (e.g., texting, using a tablet, watching a movie) will be primary rather than secondary in an L3 system.

Novice Drivers. Research has shown that novice drivers can make mistakes that experienced drivers would not make. Simply by gaining experience using the system (i.e., vehicle), the drivers become more skilled with the system and perform at a higher level. The introduction of L3 systems has the potential to change this paradigm. L3 systems may be able to correct or prevent poor decisions of the novice driver while allowing them to gain valuable experience by providing real-time feedback. The potential benefits of this may require longitudinal human factors research to fully understand.

SECTION 7. INTRODUCTION OF AUTOMATION-RELEVANT PARAMETERS

Introduction

Because vehicle automation is a market-driven system, assessing industry activity and any market barriers is an appropriate way to define concepts that are most likely to be introduced to the public. Each parameter will be described via text and summarized in a chart. The chart will serve as the basis for down-selection in the following section.

The parameters include:

- Road Facility Type
- Automated Vehicle Segregation
- Infrastructure Adaptation
- Connected Automated Operation
- Inter-Vehicle Coordination
- Speed of Travel
- Traffic Density
- Awareness of and Operation Relative to Traffic Control Devices
- Awareness of Other Vehicle Indications
- Situational Awareness
- Vehicle Maneuvers under Automated Control
- Weather Conditions
- Roadway Surface Conditions
- Driver Ability in Manual Driving
- Driver Monitoring
- Driver Task Requirement to Maintain Engagement
- Intended Duration of Automation
- Engage/Disengage Method
- Driver Engagement Timing
- Driver Training

Definition of Parameters

Road Facility Type

Definitions

Limited-access Highway: Interstate highway, high speed, absence of traffic lights, pedestrians, bicycles, merge and exit at speed

Rural Highway: High speeds, at-grade intersections, traffic lights, pedestrians, bicycles

Suburban Arterial: Moderate speeds, at-grade intersections, traffic lights, pedestrians, bicycles

City Streets: Moderate to slow speeds, frequent intersections, traffic lights, stop signs, pedestrians, bicycles Residential Streets: Slow speeds, greater prevalence of pedestrians/bikes, stop signs

Off-street (parking facilities, etc.): Typical parking lot or multi-level parking garage with pedestrians and other slow-speed vehicle traffic

Discussion

Several OEMs have prototyped and some have announced products for TJA, which provides full automation during low-speed congested highway situations. These are L2 systems. OEMs have also prototyped L2 combined lateral/longitudinal operations intended for limited-access highways. These systems typically assess lane boundaries via image processing. However, lane markings sufficient for lane detection may or may not be available on particular roadway sections.

Google has investigated the operation of automated vehicles that operate on a variety of road types. Vehicle OEMs have published research papers about situational understanding and threat detection in a variety of urban and residential environments.

Automotive suppliers have demonstrated "automated valet parking" (i.e., vehicles that drive, empty, into a parking facility to find a space and park).

Automated Vehicle Segregation

Definitions

Mixed Traffic: Automated vehicles operating in current traffic streams found on roads today

Segregated Traffic: Automated vehicles operating only on facilities restricted to automated vehicles, with physical separation from human-driven vehicles.

Discussion

While segregated traffic approaches were explored extensively in the Automated Highway Systems (AHS) program during the 1990s, all current industry activity focuses on mixed traffic.

Infrastructure Adaptation

Definitions

Adapted Infrastructure: Road facilities modified to optimally support automated vehicles by simplifying the road environment and/or enhancing the environment for sensors. This may be in the form of physically separated lanes and/or specialized road furniture/markings optimized for onboard sensors.

Non-adapted infrastructure: The current road environment including human-driven traffic with lane markings of various types and quality.

Discussion

Adaptation of the infrastructure was explored within the AHS program and other research projects that focused on "sensor-friendly highways." State Departments of Transportation (DOTs) have shown no indication that investments into automated vehicles are forthcoming.

Vehicle manufacturers do not want to wait for, or rely on, state investments and deployment – which may vary from state to state – before offering products. Therefore, design activities are focused on roads as-is. This is reflected in the prototypes and products under development (see Section 6 Level 2 and Level 3 Systems).

Connected Automated Operation

Definitions

Information derived from the Internet via cellular data communications and Global Positioning System (GPS) technology is assumed to be generally available with signal dropout a possibility in all cases.

V2V: Automation is implemented that uses IEEE 802.11p DSRC to exchange data between vehicles traveling in the vicinity of one another, including, but not limited to, the SAE J2735 Basic Safety Message.

Vehicle-to-Infrastructure (V2I): Automation uses IEEE 802.11p DSRC to receive data from the infrastructure, including curve geometry, weather/road conditions, signage, traffic signal phase and timing, and intersection geometry. Additionally, the infrastructure uses vehicle data to adapt traffic control devices and collect probe data to support area-wide traffic management.

Both: Use of both V2V and V2I.

None: Automated vehicles rely on onboard sensors and information derived from the Internet and GPS.

Discussion

V2V and V2I are the subjects of a variety of field trials. NHTSA is considering regulatory action to require DSRC on all new passenger vehicles and heavy trucks. European automakers are planning to introduce DSRC in 2015 to support a variety of safety and non-safety applications. Industry discussions have focused on connected automated systems as key to new gains in mobility and environmental benefits. For instance, the Federal Highway Administration (FHWA) is planning research into Cooperative Adaptive Cruise Control (C-ACC), which can be considered a precursor to connected automation.

For initial products, OEMs are focusing on deployment of systems that are sensor based, with the expectation that V2V and V2I data will be incorporated into situational awareness algorithms when that information becomes available.

Inter-Vehicle Coordination

Definitions

Platooning: Inter-vehicle communications are used to provide information about lead vehicle actuation to multiple automated vehicles following at close headways. The vehicular tasks are limited to following the vehicle ahead, maintaining the headway, and join/split maneuvers.

Cooperative Headway Management: Inter-vehicle communications are used to implement C-ACC, thus shortening the distance between a lead vehicle and a single following vehicle. This function can be implemented as longitudinal control alone or in combination with lateral control. C-ACC offers significant improvement in fuel economy due to the aerodynamics of shorter headways.

Individual Vehicle: "Free agent" operation; the automated vehicle coordination is typical of the operations regular drivers perform today; inter-vehicle communications do not play a role.

Discussion

Platooning was first implemented in the FHWAAHS program in the 1990s, and similar work occurred in Japan during that period. The European SARTRE program, which ran from 2009 to 2012, implemented platooning based on a business model of a lead driver being paid to drive the lead vehicle by the occupants in the following vehicles. SARTRE partners Volvo Cars and Volvo Trucks assert that their SARTRE approach to platooning is a near-term system as it relies on the lead vehicle being human-driven. The human leader can react appropriately to any anomalies on the road such that the roles of the automated vehicles are limited to car following and headway maintenance. Others have countered that asking the public to accept both automation and close-headway operations in a first-generation automation offering may not be realistic.

Additionally, the success of this concept will require significant numbers of vehicles equipped with both communications and automation capabilities. FHWA is planning light-vehicle platooning trials under its Exploratory Advanced Research program.

Cooperative Headway Management: C-ACC user acceptance research has shown that drivers quickly become comfortable with smaller gaps than those offered by current ACC systems. C-ACC experiments conducted in Europe have demonstrated traffic flow and flow stability benefits from this technique. FHWA is planning truck-based C-ACC trials under its Exploratory Advanced Research program. Peleton Aerolink is a start-up company offering retrofit C-ACC capabilities for heavy trucks. Product introduction and its value depend on sufficient penetration of V2V-equipped vehicles for customers. In the case of the trucking industry, large-truck fleets could coordinate operations such that their equipped vehicles locate one another and connect via C-ACC, which could result in fuel economy benefits.

Individual Vehicle: Extensive product-oriented industry activity focuses on individual vehicle operations.

Speed of Travel

Definitions

Low speed (0–30 mph): Applies to TJA

Higher speeds only (30–75 mph): Applies to operations conducted at typical highway speeds

Full speed range: Encompasses full range of speeds (i.e., a system that is not limited to a specific speed range)

Discussion

The speed of travel must be considered in the context of the road environment. Low-speed systems for highway use imply a vastly different set of system requirements than do low-speed systems for city driving. This parameter focuses on highway speeds.

The auto industry has initially focused on TJA because lower speeds create a situation during which the performance requirements are less challenging. TJA is also considered to provide greater utility to drivers by relieving them of the most tedious driving situation: the traffic jam. TJA will serve as a learning ground and test case for systems operating within the full speed range.

Several OEMs and Google have focused on highway speed systems that combine ACC with lane-keeping ability. The Google and GM systems operate at the full speed range.

Traffic Density

Definitions

(These are the FHWA Level of Service [LOS] definitions)**:** LOS A: Free flow

LOS B: Flow with some restrictions

LOS C: Stable flow; maneuverability and speed more restricted

LOS D: Unstable flow, temporary restrictions momentarily slow vehicle

LOS E: Unstable flow, vehicles unable to pass, temporary stoppages

LOS F: Force traffic flow condition with low speed

Discussion

Automakers are developing systems for higher-speed flowing traffic (LOSs A, B, C) during the near term and focusing on the TJA function (current product) for LOSs D, E, and F. Systems are also under development to cover the full range of traffic conditions.

Awareness of and Operation Relative to Traffic Control Devices

Definitions

Traffic signal (circular): Detection of the state of traffic signals plus the lane assignment of traffic signals for more complex intersections

Traffic signal (turn indication): Detection/understanding of turn arrows and lane assignments to which the turn arrows are applicable

Traffic circles: Detection of traffic circles, number of lanes in traffic circles, and observing proper behavior in traffic circles

Pedestrian crossings: Detection of pedestrian crossing zones and observing local rules for allowing pedestrians to cross

Lane restriction indications: Detecting and observing turn-only lanes and/or restrictions to certain vehicle types; this includes electronic message signs

Work zones: Detecting work zones, special signed instructions, and lane designations via traffic cones or other markers.

One-way street signs: Detecting and observing signage indicating one-way streets

Yield signs: Detecting and observing yield signs

Stop signs: Detecting and observing stop signs and behaving according to appropriate order of precedence

Speed signs: Detecting speed signs and adjusting speed accordingly, including electronic speed signs

Discussion

To some degree, automated vehicle systems will use highly accurate map data. However, the vehicles must be capable of detecting actual traffic control devices on the road to avoid a violation.

Traffic Sign Detection is a product that has been available in Europe since approximately 2010. Systems that will work in the U.S. are still under development because the U.S. does not follow international sign conventions. Detection of traffic signals has been implemented on prototypes; the robustness of detection is the focus of further technical development. Research has also been performed on operation in traffic circles.

Awareness of Other Vehicle Indications

Definitions

Brake lights: Detection of brake lights on a lead vehicle as a redundant indication of deceleration (i.e., a backup to sensors and communication)

Turn signals: Detection of turn signals on lead and/or adjacent vehicles as a means of assessing the intentions of other drivers

Discussion

No information was found in the literature about prototypes that detect either brake lights or turn signals.

Situational Awareness

Definitions:

Vehicles: Nearby vehicles Motorcycles: Nearby motorcycles

Road condition: Detection of surface traction condition, potholes, etc.

Road debris: Detection of debris on the road (e.g., tire/animal carcasses, items that may have fallen off other vehicles)

Pedestrians: Pedestrians relevant to the planned travel path, particularly in right-turn movements made across crosswalks when the pedestrian is traveling in the same direction (the pedestrian may be in the driver's blind spot)

Bicyclists: Nearby bicyclists, particularly in right-turn movements made across crosswalks when the bicyclist is traveling in the same direction (the bicyclist may be in the driver's blind spot)

Animals: Awareness of animals on the road and those on the roadside who may bolt across the road

Discussion

Prototypes and products designed for current highway operations are aware of other vehicles, motorcycles, and pedestrians based on active safety offerings that have been available for several years.

The road condition can be derived from traction control systems. However, preview information is more important, though this is not possible with onboard sensors. Information obtained from lead vehicles using V2V communication will be helpful in this situation.

Some road debris may be detectable by current systems. Again, V2V communication systems will be helpful in this situation.

Prototypes for intersection crash avoidance[23] and urban automated driving have been designed to detect bicyclists.

Tier One suppliers are now working on systems designed to detect large animals[26].

Vehicle Maneuvers under Automated Control

Definitions

Stay in Original Lane Only: Cruising at set speed; lane markings sufficient for lane detection may or may not be available

Lane Change: Monitor adjacent lane for adequate gap; signal and execute lane change; lane markings sufficient for lane detection may or may not be available

Freeway-to-Freeway Interchange: Maneuver to proper lane for merging onto the desired freeway and direction; merge into traffic on new freeway

Diverge from Freeway to Surface Street: Signal for exit from freeway; detect and understand configuration at end of departure ramp; select proper lane and observe yield/stop/signal information

Merge into Traffic: Merge into freeway traffic from surface street ramp with awareness of merging distance available

Left Turn across Traffic: Make a left turn with awareness of any crossing pedestrians or bicyclists; awareness of any oncoming vehicles that may come into sensor view during the maneuver

Right Turn: Make a right turn with awareness of any crossing pedestrians or bicyclists who have the right of way

Stop/Start at Traffic Signal: Awareness of traffic signal state for travel lane; maintain proper gap behind lead vehicle while accelerating on green signal

Stop-Controlled Intersections: Awareness of stop sign and configuration (two-way, four-way); awareness of other vehicles at intersection and which vehicles have right of way; observance of order of precedence when multiple vehicles are at the intersection

Traffic Circles: Awareness of traffic circle configuration, including number of lanes; observance of proper lane behavior; awareness of vehicles already in the traffic circle; awareness of right of way for each vehicle; enter and depart traffic circle following local rules

Work Zones: Detection of lane boundaries (possibly unconventional) and/or stop indications; maneuver through work zone with awareness of other traffic and work zone speed limits; merge into adjacent lane in lane-drop situations

Respond to Emergency Vehicles: Awareness of emergency vehicle and location relative to other vehicles; determine need to move out of the path of the emergency vehicle; determine safe and legal options to move out of the emergency vehicle path and execute the ideal maneuver

Yield as Appropriate: Awareness of yield signs and road rules for yielding in the absence of signs; awareness of other vehicles and their right-of-way status; proceed in accordance with right of way

Discussion

The vehicle industry is developing prototypes designed for lane changes, freeway-to-freeway merging, handling traffic circles, and negotiating work zones.

Researchers focusing on automated urban driving have developed prototypes capable of making left/right turns and responding to traffic signals and yield/stop situations.

Activity is either focused on freeways or urban driving, not both. Therefore, no activity has been detected to accomplish divergence from a freeway onto a surface street or vice-versa.

No activity has been detected for automated vehicles responding to emergency vehicles.

Weather Conditions

Definitions

No Adverse Conditions: Dry, clear

Rain: Adapting speed as needed due to wet pavement and visibility

Sleet: Adapting speed as needed due to pavement condition and visibility; alerting driver when automated control is no longer possible

Snow: Adapting speed as needed due to pavement condition and visibility; alerting driver when automated control is no longer possible (e.g., due to obscured lane markings)

Fog: Adapting speed as needed due to visibility; alerting driver when automated control is no longer possible Other (Smog, Smoke, Sand/Dust, Crosswind, Hail):Adapting speed as needed due to visibility; alerting driver when automated control is no longer possible; adapting steering as needed in crosswinds

All Conditions: Automated vehicle can operate at minimum performance standards regardless of weather

Discussion

Regarding the full range of weather conditions, systems may not be able to determine when the weather falls outside of the verified performance envelope.

First-generation automated vehicle systems will most likely disable under heavy rain (poor visibility) conditions using rain detection systems already available on production cars. The ability to transfer control back to the driver during these cases (e.g., instantaneous heavy rain in a thunderstorm) will be critical.

Current systems are capable of operating during at least moderate rainfall. Active safety systems (i.e., the technology basis for automation) have the ability to "see" through fog and other visual obscurations. However, the majority of current work focuses on moderate-to-good weather conditions, leaving inclement weather operation to later-generation systems.

Roadway Surface Conditions

Definitions

While metrics for roadway surface conditions are important in a requirements document, they are not needed for this high-level ConOps document.

Discussion

Similar to discussion about weather parameters (above), current work focuses on ideal pavement conditions. Operation during degraded conditions (e.g., snow/ice/oil) will be left to future-generation systems.

Driver Ability in Manual Driving

Definitions

Novice: A learning driver or one with less than one year of experience

Experienced: A driver with one or more years of driving experience

Impaired Due to Age or Disability: Drivers whose driving skills have started to degrade due to age; drivers whose cognitive and/or motor skills are degraded due to a medical condition

Discussion

Current work focuses on the average driver (i.e., an experienced driver).

Research has shown that novice drivers (e.g., teens) can make errors due to lack of skill and/or impulsive behavior. Current automation prototypes can support novice drivers, and driver interfaces could be developed to coach novice drivers. Coaching systems have been developed for some insurance products that could serve as a starting point.

Driver impairment can result from drowsiness; this condition is addressed to some degree by current driver fatigue countermeasures offered by several automakers. Automated systems that allow persons incapable of driving to "be driven" are being explored by the industry. The Emergency Stopping Assistant developed by BMW provides safe stopping when a driver becomes unresponsive.

Driver Monitoring

Definitions

Yes: Systems are implemented that monitor some or all of lane-keeping stability, pedal application, steering inputs, manipulation of other controls, seating position, eye-blink rate, and gaze (on or off road)

No: No monitoring at all

Discussion

Driver fatigue systems currently available on the market monitor driver activity in a basic way (e.g., by noting lane-keeping stability, pedal application, steering inputs, and manipulation of other controls [e.g., radio and climate control]). However, as noted by HAVEit researchers, automated driving reduces or eliminates these data elements.

A key driver-monitoring technology is "hands on wheel" detection achieved through counter-torque detection or capacitive touch sensing. Several LKASs on the market now implement some form of this technology.

In terms of driver visual attention, driving distraction activity has led to research into driver-monitoring techniques. However, the auto industry has not yet extensively deployed driver monitoring. Some industry executives have made public statements that driver monitoring is essential for L2 automated driving.

Based on experiments, the HAVEit researchers concluded that the most promising measures for driver state assessment during highly automated driving were vision-based measures that analyze the driver's eye closure behavior for drowsiness detection and analyze the driver's head or gaze direction for distraction.

Overall, driver monitoring is an active discussion topic within the vehicle industry. However, no public exhibitions of automated driving combined with driver monitoring have occurred at this point. Based on information released by the industry, TJA products that have been announced do not implement driver monitoring.

Driver Task Requirement to Maintain Engagement

Definitions

Yes: Driver must take an action of some sort at defined intervals for the automated system to continue operation

No: The driver needs to take no action for the automated system to continue operation

Discussion

A task requirement to maintain driver engagement could assume a variety of forms in terms of the action itself, time periods, and specific conditions. A driver task requirement may work in tandem with driver monitoring as well (i.e., the vehicle may prompt the driver to perform a task only after first determining via driver monitoring that the driver's visual attention to the driving scene is inadequate).

For the TJA system on the 2014 Mercedes S-Class, the driver is expected to be engaged, with his or her hands placed on the steering wheel. If the driver is not touching the steering wheel, the system will disable after a few seconds.

Press materials from other automakers do not comment about the possibility of a driver task requirement. Press reports from the Audi TJA demonstration made during the 2013 Consumer Electronics Show[27] note that the system offers video entertainment to the driver once the system assumes control of driving.

Intended Duration of Automation

Definitions

Short: Less than 1 minute

Medium: 1–10 minutes

Extended: 10–30 minutes

Long: Greater than 30 minutes

Discussion

There is no indication in the literature that the vehicle industry and researchers are defining their systems in terms of duration of automation.

Engage/Disengage Method

Definitions

System Request: The system makes a request to the driver to engage or disengage

Driver-Initiated: The driver requests the system to engage or disengage from driving

Both: Both options are implemented

Forced Disengage/Failure: The system requests the driver to resume control due to inadequate road conditions or a system failure; in the case of a non-responsive driver, the system brings the vehicle to a (relatively) safe state

Discussion

Published information about TJA systems indicates that the vehicle system will inform the driver when the conditions have been met for TJA operation to commence; the driver then decides whether or not to engage the system.

When a TJA system detects a condition requiring driver control, the system will request that the driver reassume control. In the case of the Audi prototype, if the driver does not respond, the vehicle will tap the brakes after five seconds. If there is still no response, the car begins to gradually brake to a stop within the next 10 seconds and activates the hazard lights. The specific actions of TJA systems developed by other OEMs have not been published.

The European HAVEit project focused on the human factors design of the transitions between lower and higher degrees of automation, including development of a progressive step-by-step approach to transfer the driving task from the automated system to the driver. When returning control to the driver, the interaction began early in the event chain (i.e., a few seconds before a potentially critical situation occurred). According to the research report, "the system brought the driver back into the loop in advance of the critical situation and provided him or her with the optimum level of automation and assistance needed in critical situations."

Driver Engagement Timing

Definitions

Engagement timing refers to the amount of time the system allows the driver to re-engage operations after a period of automated driving.

Short: Less than 3 seconds

Medium: 3–10 seconds

Long: 10–30 seconds

Extended: Greater than 30 seconds

Discussion

Engagement timing is critical to assessing the safety of automated driving concepts and relates to the engagement methods, any driver task requirements, and the potential need for driver monitoring. The driver engagement time could vary by the driver type (e.g., novice, experienced, impaired). In the case of a person incapable of driving, driver engagement time becomes infinite (i.e., the system becomes an L4 system and must handle all aspects of driving at all times).

Public information about TJA products and prototypes does not list specific engagement timing. Information about the Audi prototype indicates a 15-second engagement period.

Assumptions about engagement timing relate to the risk of non-engagement. Information about the Audi prototype indicates that its TJA system may bring the car to a halt, and activates hazard lights, when the driver is unresponsive. This operation is relatively safer than proceeding when the system is incapable of driving the vehicle.

The European HAVEit project defined two "Minimum Risk Maneuvers" during this type of situation:

- Bring the vehicle to a stop on-road
- Change lanes as needed to bring the vehicle to a stop on the shoulder of the road

Driver Training

Definitions

None: No driver training; the driver discovers system operation and forms his/her mental model of system operation

Minimal: The driver is provided information (e.g., a video shown at the car dealership) before driving the vehicle for the first time

Substantial: Before driving an automated vehicle, a special driving certification based on training must be obtained.

Discussion

No information was discovered that indicates automakers are designing systems that require training.

The Nevada autonomous driving law[28] requires that drivers obtain a special endorsement on their driver's licenses.

SECTION 8. SELECTED AUTOMATION CONCEPTS FOR HUMAN-MACHINE INTERFACE EVALUATION

Down-selected Parameters

In *Section 7 Introduction of Automation-Relevant Parameters*, a broad set of parameters was identified for the L2 and L3 automated vehicle concepts. From that larger set, a down-selected set was identified based on industry activity in research and development. Table 8-1 provides this smaller set by Parameter and High-ranking Modality.

Selected Automation Concepts for Human-Machine Interface Evaluation

The following factors were identified as common to all automation concepts:

- Highway
- Mixed traffic
- Infrastructure as-is

- Only cellular and GPS connectivity
- No vehicle coordination (individual vehicle operation)
- Awareness of vehicles, motorcycles, pedestrians
- Dry and wet weather/road conditions
- Novice and experienced drivers
- Engage/disengage by driver or system

The following figure (8-1) represents a model for investigating the relevant operational parameters for each chosen concept. Each channel represents a different set of conditions for evaluation. There are 32 channels total, thus ensuring a thorough investigation of each concept under each of the conditions presented.

Table 8-1. Vehicle Automation Parameters with High Industry Activity and Likely Near- or Mid-term Introduction

Parameter	High-ranking Modality
Road Facility Type	Limited-access highway
Automated Vehicle Segregation	None (mixed traffic)
Infrastructure Adaptation	None
Connected Automated Operation	Cellular/GPS only
Inter-Vehicle Coordination	Platooning (research)
Inter-Vehicle Coordination	No coordination (individual vehicle; industry)
Speed of Travel	0–30 mph (earliest market timing)
Speed of Travel	30–75 mph (near term)
Traffic Density	All LOSs
Awareness of and Operation Relative to Traffic Control Devices	Speed signs
Awareness of Other Vehicle Indications	None
Situational Awareness	Vehicles, motorcycles, pedestrians
Vehicle Maneuvers Under Automated Control	Stay in original lane
Weather Conditions	Dry, clear, rain
Roadway Surface Conditions	Dry, wet
Driver Ability in Manual Driving	Novice, experienced
Driver Monitoring	Both monitored and unmonitored
Driver Task Requirement to Maintain Engagement	Yes and no
Intended Duration of Automation	All
Engage/Disengage Method	Initiated by driver/system/both
Driver Engagement Timing	All
Driver Training	None

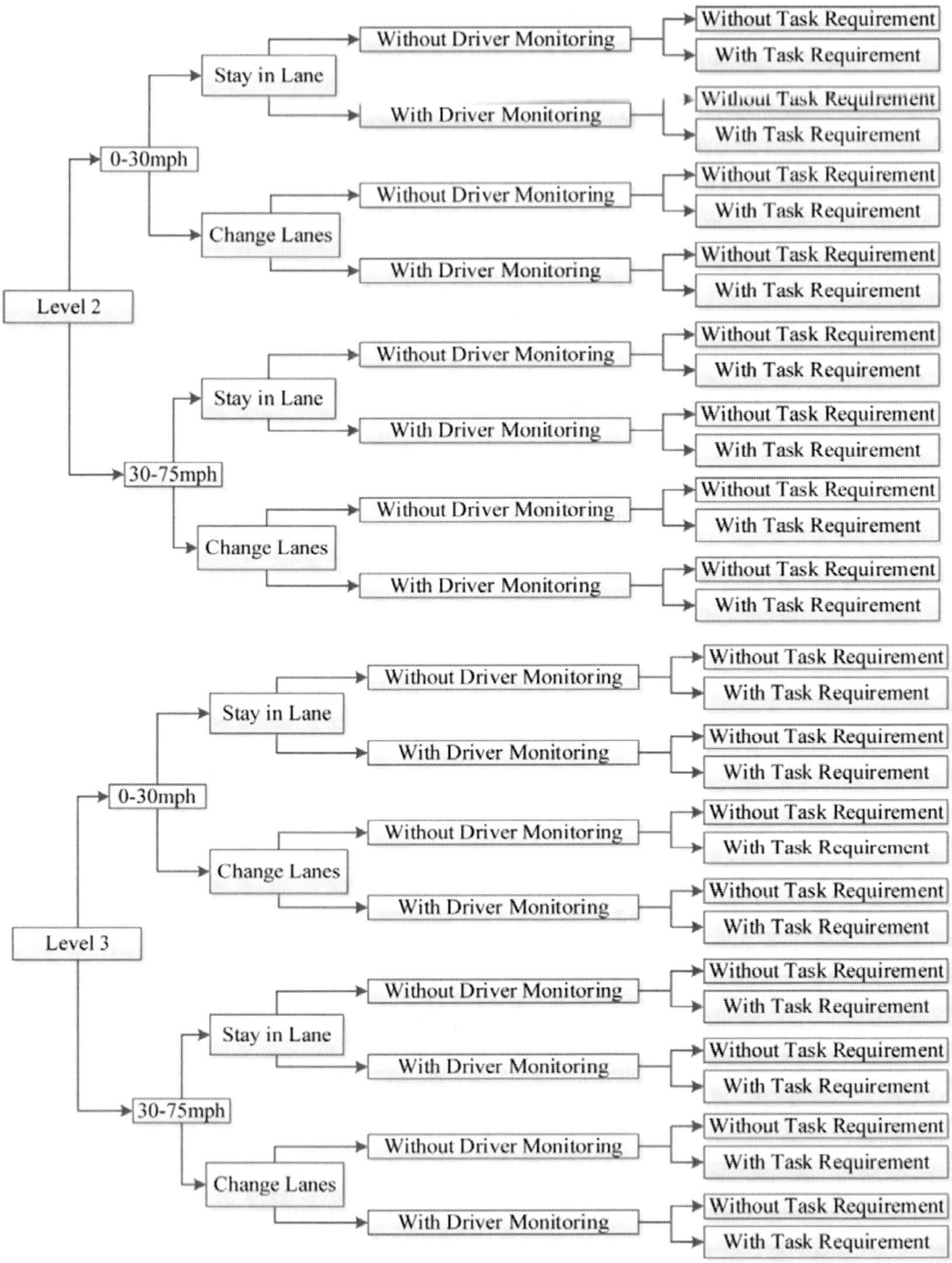

Figure 8-1. Relevant Operational Parameter Investigation Model for L2 and L3 Systems (Source: Author).

SECTION 9. OPERATIONAL SCENARIOS

This ConOps document identifies the existing systems, the proposed systems, and the current state of vehicle automation efforts (in combination with the Literature Review document). This ConOps may also be used to inform subsequent research about L2 and L3 systems. Operational scenarios are descriptions of vehicle operations, which include operator actions, that define a specific transport situation. For the operational scenarios, a specific literature search was conducted that identified analyses of data sets providing insight into the most common pre-crash indicators (or scenarios, depending on the study). Pre-crash indicators (or scenarios) are the kinematic attributes of the vehicle prior to, or operator actions leading to, a crash or near-crash.

More than 20 reports or journal articles about transportation data sets were reviewed to identify the key attributes and characteristics of pre-crash scenarios. Of these, six studies stated findings related to the vehicle kinematics or operator actions associated with negative outcomes. The following six studies were evaluated for pre-crash scenarios:

- Integrated Vehicle-Based Safety Systems (IVBSS) Light Vehicle Extended Pilot Test Summary Report (IVBSS)[6]
- Heavy Vehicle-Light Vehicle Interactions Data Collections and Countermeasure Research Project Phase 1 – Preliminary Analysis of Data Collected in the Drowsy Driver Warning System Field Operational Test: Task 5, Preliminary Analysis of Drowsy Driver Warning System Field Operational Test Data, DTNH22-00-C-07007, Task Order #21 (34 Truck)[5]
- Frequency of Target Crashes for IntelliDrive Safety Systems, DOT HS 811 381 (VOLPE, GES)[3]
- The 100-Car Naturalistic Driving Study, Phase II – Results of the 100-Car Field Experiment, DOT HS 810 593 (100 Car)[2]
- Naturalistic assessment of novice teenage crash experience (40 Teen)[25]
- Investigating Critical Incidents, Driver Re-start Period, Sleep Quantity, and Crash Countermeasures in Commercial Vehicle Operations Using Naturalistic Data Collection (8 Truck)[4]

The OVI system is concerned with both the operator and the vehicle. Thus, all applicable attributes were collected regardless of which subsystem was the most appropriate. The resulting relationships of attributes were mapped with and between studies. Figure 9-1 provides this map and indicates through line thickness which attributes were most common. It should be noted that the parenthetical names of the aforementioned studies are used in the figure (e.g., IVBSS, 100 Car). Often, a different term or phrase was used to describe a similar pre-crash scenario (indicator) across studies. To simplify the mapping where possible, multiple attributes were combined where similar definitions existed.

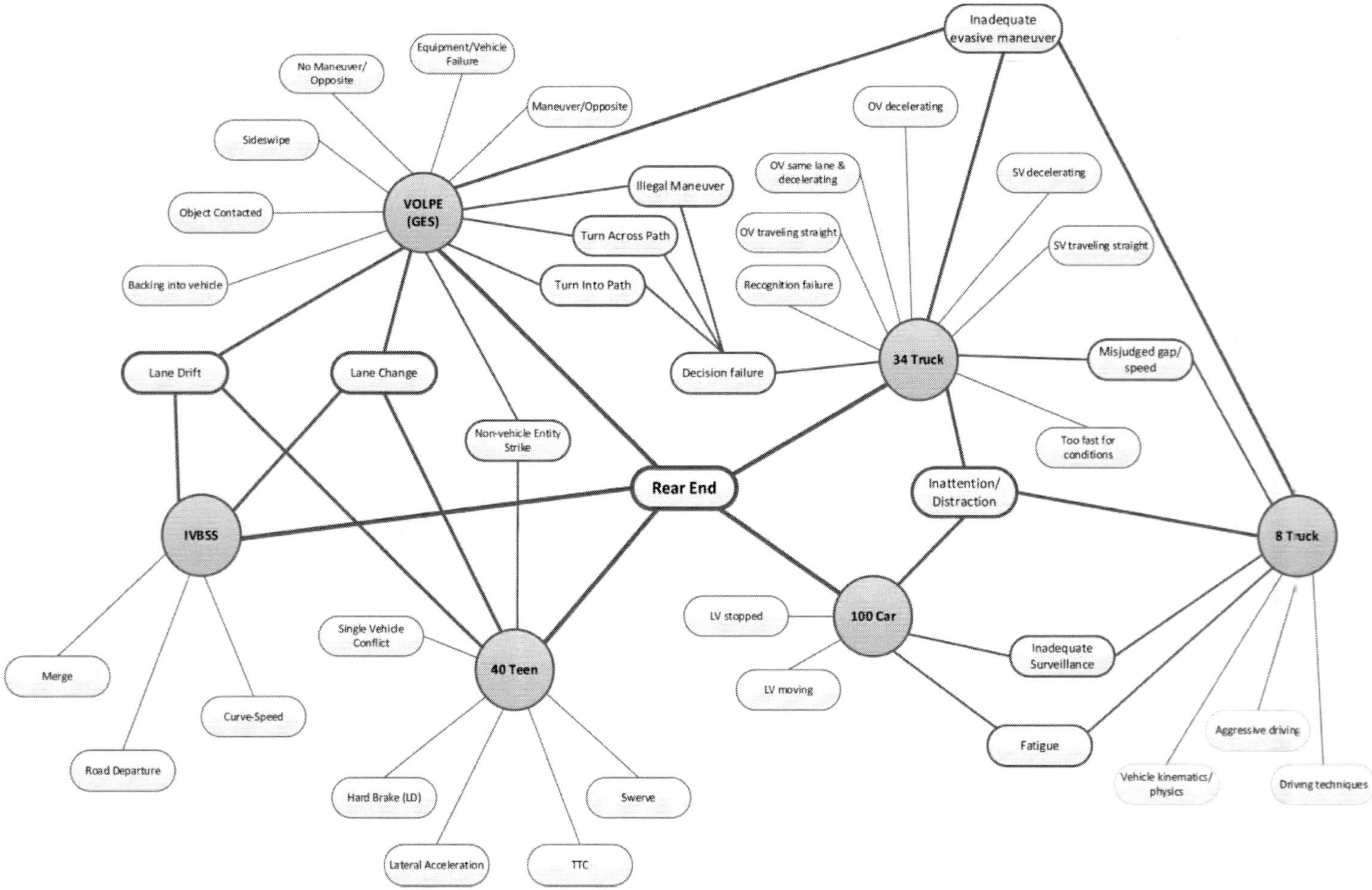

Note on abbreviations: lane drift (LD); time-to-collision (TTC); lead vehicle (LV); subject vehicle (SV); other vehicle (OV).

Figure 9-1. Pre-crash Scenario Mapping (Source: Author).

There are some inherent biases in this evaluation. For instance, the 34-Truck data set was collected from a study evaluating a drowsy driver warning system, which may be one reason fatigue was not specifically mentioned in the report. While this type of bias exists, the objective of this evaluation was to aggregate the findings of multiple data sets (five of the six data sets are derived from naturalistic studies) to indicate potential areas of focus for research. The 10 most common pre-crash indicators/scenarios found during this effort were as follows (the number of studies mentioning these indicators/scenarios are noted in parentheses):

1. Rear-end (5)
2. Lane Drift (3)
3. Lane Change (3)
4. Inattention/Distraction (3)
5. Inadequate Evasive Maneuver (3)
6. Inadequate Surveillance/Eyes off Roadway (3)
7. Fatigue (2)
8. Conflict with Non-vehicle (2)
9. Decision Failure (2)
10. Misjudge Speed or Gap (2)

Purpose of Scenarios

Scenarios serve to place concepts into specific environments to flesh out their functional characteristics in the real world. The following are four categories of scenario attributes intended to replicate the majority of situations anticipated for follow-on testing and evaluation.

Roadway environments:

- Rural interstate: I-81 Blacksburg via I-66 to Washington, D.C.
- Urban interstate: D.C. Beltway
- Rural highway (two-lane): Blue Ridge Parkway
- Urban streets: Downtown Detroit
- Residential streets: Typical neighborhood Type of traffic:
- Congested
- Free-flowing

Maneuvers:

- Cruising (moving at set speed unimpeded by traffic)
- Vehicle following
- Lane change
- Freeway-to-freeway interchanges
- Negotiating work zones
- Normal stop and go
- Reacting to cut-ins by other traffic

- Detecting/obeying turn-only lanes
- Observing proper right-of-way for pedestrians and bicyclists
- Detecting/avoiding non-vehicle obstacles

Traffic control devices:

- Stop/yield signs
- Speed signs
- Turn restrictions
- Lane restrictions
- One-way streets
- Ramp meters
- Traffic signals

These scenarios are not meant to be inclusive. Instead, they are designed to provide guidance in combination with other factors for developing experimental designs. Other factors include the identified pre-crash scenarios and selected automation concepts for human-machine interface evaluation. Table 9-1 maps these attributes across roadway types.

Table 9-1. Roadway to Attribute Mapping

Attributes	Rural Interstate	Urban Interstate	Rural Highway	Urban Streets	Residential Streets
Cruising	√	√	√	√	√
Vehicle following	√	√	√	√	√
Lane change	√	√		√	
Freeway-to-freeway	√	√			
Negotiating work zones	√	√	√	√	√
Normal stop and go	√	√	√	√	√
Reacting to cut-ins	√	√		√	
Detecting/obeying turn-only lanes	√	√	√	√	√
Interactions with pedestrians and bicyclists			√	√	√
Detecting/avoiding non-vehicle obstacles	√	√	√	√	√
Stop/yield signs			√	√	√
Speed signs	√	√	√	√	√
Turn restrictions			√	√	√
Lane restrictions	√	√		√	
One-way streets				√	√
Ramp meters	√	√			
Traffic signals			√	√	√

It is important to note that the scenarios identified in this section along with the parameters identified in Section 10 and all other information presented herein are designed to better specify the domain of system operations with respect to L2 and L3 automated vehicles. There is a bias towards those technologies expected to arrive first in the marketplace. To the extent possible, the information provided in this document is intended to inform the reader of the operational framework for these vehicles and to provide focus for proposed follow-on research.

SECTION 10. SUMMARY OF IMPACTS

Implementation of the proposed L2/L3 systems may have broad impacts on all system stakeholders. The following sections identify potential operational impacts, organizational impacts, and developmental impacts that should be considered for development, testing, and evaluation of the L2/L3 systems.

Operational Impacts

The OVI system is a general description of L2/L3 systems expected to be developed and implemented in multiple ways. Each manufacturer will develop the system as they see fit. Thus, there is some uncertainty regarding the specific types of L2/L3 systems that will be initially produced. This ConOps document is meant to be general in nature so as to be applicable to all system types. Therefore, consideration of potential operational impacts could include the following:

- Methods for controlling the system
- Information presented to operator
- Human-machine interface efficacy
- Human factors issues (disadvantages and limitations of various human-machine interface approaches)
- Modes of operation per system
- System malfunction/failure transition
- Novice user barriers

This list is not meant to be exclusive. It instead offers guidance as the development and testing phases continue. It is expected as systems finish development and move to production, information about specific systems will be available for a complete review of operational impacts.The OEMs of L2/L3 systems are expected to be organizations with experience in equipment development, testing, and evaluation. While implementation of these systems may qualify as disruptive to the operator (i.e., technology sufficiently different such that there is a learning curve for the operator), it is not expected that these systems will be disruptive to OEM operations. It should be noted that issues such as manufacturing re-engineering, component redesign, and market impacts are beyond the scope of this ConOps document.

Organizational Impacts

Currently, NHTSA is responsible for implementing and enforcing the National Traffic and Motor Vehicle Safety Act of 1966, as amended, 49 U.S.C. Chapter 301 (the Vehicle Safety Act), and certain other laws relating to motor vehicle safety.[29] Under that authority, NHTSA issues and enforces Federal motor vehicle safety standards (FMVSS) that apply to motor vehicles and to certain items of motor vehicle equipment. The Vehicle Safety Act requires that motor vehicles and regulated items of motor vehicle equipment manufactured for sale in the United States be certified to comply with all applicable FMVSS. Type approval is not required for motor vehicles and motor vehicle equipment sold in the United States. NHTSA does not issue type approval certifications and does not certify any motor vehicles or motor vehicle equipment as complying with applicable FMVSS. Instead, in accordance with 49 U.S.C. 30115, a "self-certification" process is in place, which requires the manufacturer to certify the vehicle or equipment item as complying with the applicable FMVSS.[29] In general, it is possible that introduction of L2/L3 systems may require standards to ensure aftermarket systems can be integrated post-production without adverse effects to the OVI system. Consideration should be given to the nature in which such standards should exist.

Impacts during Development

The development and implementation of L2/L3 systems may require involvement from multiple stakeholders, including government, vehicle manufacturers, system developers, and other support organizations. The development process requires consideration of the integration of these various partners. Furthermore, collaboration during the early stages of development could benefit from discussion of functional standards for the L2/L3 systems to evaluate the potential impacts of a consistent experience to the operator with regards to functionality and expectation.

INDEX

C

D

H

I

J

K

L

M

N

O

P

Q

R

S

T

U

V

W

Y